The Wisdom of the Body

ALSO BY ERIK SHONSTROM

Wild Curiosity: How to Unleash Creativity and Encourage Lifelong Wondering

The Indoor Epidemic: How Parents, Teachers, and Kids Can Start an Outdoor Revolution

The Wisdom of the Body

What Embodied Cognition Can Teach Us about Learning, Human Development, and Ourselves

Erik Shonstrom

ROWMAN & LITTLEFIELD
Lanham • Boulder • New York • London

Published by Rowman & Littlefield
An imprint of The Rowman & Littlefield Publishing Group, Inc.
4501 Forbes Boulevard, Suite 200, Lanham, Maryland 20706
www.rowman.com

6 Tinworth Street, London SE11 5AL

Copyright © 2020 by Erik Shonstrom

All rights reserved. No part of this book may be reproduced in any form or by any electronic or mechanical means, including information storage and retrieval systems, without written permission from the publisher, except by a reviewer who may quote passages in a review.

British Library Cataloguing in Publication Information Available

Library of Congress Cataloging-in-Publication Data

Names: Shonstrom, Erik, author.
Title: The wisdom of the body : what embodied cognition can teach us about learning, human development, and ourselves / Erik Shonstrom.
Description: Lanham : Rowman & Littlefield, [2020] | Includes bibliographical references. | Summary: "The Wisdom of the Body uses embodied cognition studies to challenge conventional approaches to learning, and offers compelling examples of how an embodied approach to education could revolutionize schools"—Provided by publisher.
Identifiers: LCCN 2019042110 (print) | LCCN 2019042111 (ebook) | ISBN 9781475840643 (cloth) | ISBN 9781475840650 (paperback) | ISBN 9781475840667 (epub)
Subjects: LCSH: Cognitive learning—Physiological aspects.
Classification: LCC LB1062 .S485 2020 (print) | LCC LB1062 (ebook) | DDC 371.39—dc23
LC record available at https://lccn.loc.gov/2019042110
LC ebook record available at https://lccn.loc.gov/2019042111

For my dad.

Contents

Acknowledgments		ix
Author's Note		xi
Introduction: San Antonio Blues		xiii
1	Bodies of Knowledge	1
2	A Brief History of a Good Idea at the Time	17
3	Evolutionary Baggage Claim	27
4	Got Dopamine?	45
5	The DeMoN in Our Heads	53
6	Gandalf of Brooklyn	63
7	Bodies of Literature	75
8	Prose in Repose	87
9	Deconstructive Criticism	97
10	Monkey See, Monkey Do	107
11	Gut Feelings	115
12	Virtual Race	129
13	Anything Boys Can Do	143
14	Golden Oldies	153
15	Dionysian Embodiment	161
16	The Jumper	167
Appendix: A Practical Guide to Embodied Education		169
Bibliography		173
About the Author		181

Acknowledgments

Big time thanks to my editor at Rowman & Littlefield, Sarah Jubar, for using her sharp eye and patient hand to guide this book to the finish line. I'm also grateful for the team at R&L that helped with the final stages: Carlie Wall, Kira Hall, Andrew Yoder, and Tom Koerner.

For the cheap price of a cup of coffee or two, the following people read chunks of the original manuscript and provided advice, encouragement, and most importantly, corrections: Kathy Seiler, Cyndi Brandenburg, Mike Kelly, Rebecca Mills, and Herbert Petri. They have my gratitude. Special thanks to Dave Mills, who took time away from a busy schedule to explain about three hundred years of philosophy to me. Thanks, man.

Extra special thanks and seam-busting love to Vivien Enriquez, who educated me on the gut microbiome and provided incredible insight and feedback. Thanks also to Viv for getting me behind the scenes at an aquarium so I could play with an octopus—so cool. Love you, kid.

For conversations about embodiment and teaching, thanks to Tim Brookes, Kristin Novotny, Patricia DeRocher, and Josh Meyer. Thanks to Kristin Wolf for lending the Abram book—that was a game changer. Thanks to my buddy Adam Rosenblatt for hanging out, keeping me interested in my own work, and inspiring me to think more radically about bodies and learning.

Thanks to Matthew Costello for chatting with me and explaining embodiment and aging.

Profound gratitude is due to Pamela Formica, whose patience, wisdom, and kindness will not be soon forgotten.

For all the students that have ever taken my course, *Bodies*, thanks for unwittingly being guinea pigs for many of the ideas in this book.

Love and gratitude to Finn Shonstrom, who lives so joyfully through his body that it inspires me to be a better teacher, writer, father—and also to give up on my inbox once in a while and go jump in the lake. You're totally blasto, Sako.

In the course of writing this book, one of the things I've realized—on a personal level—is how lucky I am to be in love and share life with someone who engages me on every level of my senses. None of this—the book, sure, but also life, family, joy, meaning and purpose—would exist without my wife, Cindy. Love you.

Author's Note

The field of embodied cognition is vast. No individual could claim expertise in all the different disciplines it touches: cognitive science, psychology, biology, evolutionary studies, neuroscience, social psychology, and philosophy—just to name a few. Thus writing an approachable book exploring embodied cognition through the lens of education is challenging. In addition, a considerable amount of research is highly theoretical and specific, addressing a minute pinprick of the embodied cognition landscape.

I have tried to do justice to technical complexity while also making the ideas generally applicable and digestible. To bring this multifaceted field into focus, I use real-world examples whenever possible. There is a mountain of embodied cognition studies out there; I've been as diligent as I can in attributing work appropriately and acknowledging the researchers and authors responsible for developing this fascinating field. Obviously, any omissions or misrepresentations are solely my fault.

Embodied cognition (broadly construed) posits that human thinking, feeling, and learning happens bodily in concert with our mind. A natural question would be "So what?" I hope in some small way this book explores the conversations resulting from that query in a manner of interest to teachers and educators but also to anyone interested in how we learn, in the way memories form and inform decisions, and in the relationship between our sensory experience of the world and our behavior. And maybe—just maybe—this book can shed a little light on how embodiment helps us understand who we are.

Introduction

San Antonio Blues

I wore an enormous purple fleece jacket that made me look like the wobbly, rotund offspring of Barney, the purple dinosaur. I'd tuck my infant son, Finn, into a sling that held his little body close to mine. I'd zipper up and he'd be cocooned against me, ensconced in the fuzzy fabric of the coat. Despite his initial whimpers and protests, I'd begin crunching through the snow, walking out into the fields and forests around our home. He'd fall asleep quickly, wrapped in the womb-like sling across my chest. His body would jostle gently as I marched along, breathing heavily. Winter crows would caw overhead.

Those moments were some of the happiest of my life. Despite the fact that it was ten years ago and, like most new parents, I was sleep deprived, unshowered, and exhausted, the walks I took with Finn curled against me like a hibernating squirrel are as clear in my mind as though they just happened this morning. I can remember everything: the way his weight would begin to make one side of my neck sore; the lengths I'd go to avoid chatty neighbors walking their dog who might stop me for conversation and wake him up; the way my footsteps and breathing would fall into a shuffling rhythm with the gentle sway of his body; the little *gah* sound he'd make when he finally woke; the clear blue of his eyes as I unzipped him and he poked his head out into the cold, fresh air, ready for the next moments of his life.

No matter what else was going on in those days, I felt existentially complete and supremely contented during those walks. It was actually a deeply troubling time for our family; illness and financial insecurity had hit us hard and brought with it boatloads of stress. It was only during those long walks in that ridiculously oversized jacket that I felt present and within myself.

Fast-forward ten years.

I was waiting to board my flight in an hour or two. The San Antonio airport sits just outside the city. Small and efficient, it's not the tangled snarl of Newark or the endless concourse-of-death of Detroit, nor is it the massive, mind-numbing sprawl of Dallas–Fort Worth. The coffee shop sits happily right beside the TSA body-cavity-search stations. After a brief conversation with a missionary headed back to Panama as we both put our shoes and belts back on, in which we made small talk about the weather (he thought the one year he'd spent in West Virginia was "the coldest" thing he'd ever experienced—as a Vermonter, I could only chuckle inwardly and feel superior after several decades of freezing my ass off, a small but hard-won early morning victory meaning absolutely nothing), I grabbed a coffee from the kiosk and went to sit and wait for my flight. It was just after six a.m.

I had every reason to be happy. I'd just spent the past few days with my daughter Vivien at a STEM conference at which she'd presented her graduate research. Seeing her, all grown up and engaged in the pell-mell world of scientific research, is the sort of thing a parent dreams about: a happy, well-adjusted and brilliant (if I say so myself) kid on the path to a fulfilling life. She'd given a talk on the symbiotic relationship between bobtail squid and phosphorescent bacteria in which she demonstrated a keen intellect, relaxed charm, and humbly compelling thesis. I should've been Dick-Van-Dykeing my way through the airport, singing and dancing and happy.

Only I wasn't.

I felt pretty rotten. Physically I was OK—it was my emotions that were out of whack. I felt anxious and depressed, my mind caught in shitty, cyclical patterns of worry and self-criticism. It was the typical stuff of general neuroticism and self-loathing: I'm not good enough, smart enough. I'm a bad parent, useless teacher, crappy writer. No one really understands me. The usual banal insecurities of middle age—anxieties about money and success and reckoning with mortality. I was spiraling downward into a pretty negative state.

Add to that I was at the airport, a crossroads of humanity if ever there was one. Rather than seeing families on their way to well-earned vacations, travelers journeying to far-flung destinations across the world, and business-people necessarily greasing the wheels of global commerce and knitting together the fabric of our economy, I saw a faceless and shapeless mass intent on consuming the planet down to the bones, stuffed to the gills with Chick-fil-A and Dunkin Donuts, riding off into an apocalyptic sunrise in fossil-fuel-burning torpedoes of death. I was one part self-denigration, two parts asshole.

I couldn't figure out what was wrong with me. Even my morning coffee—usually my favorite moment of the day—seemed bitter and unrewarding. I stood up and started walking. I wandered a bit in the newsstand, but every newspaper and magazine and book just revved up my feelings of impending doom mixed with personal insufficiency more. Finally I just started striding across the waxed floors of the airport. Walking fast, I went to the very end of the terminal, executed a tight hairpin, and started trekking back the other way. I loped back and forth across the San Antonio airport, stomach grumbling, arms swinging. I'd only been there two nights; my backpack was light. Early morning sun shone through the windows.

It happened imperceptibly at first, but then, even as scattered and twingy as I felt, I began to notice something. I was calming down. Feeling a bit better. The more steps I took, the more the dark monsters of my psyche receded. I saw a toddler with her parents. She was capering about in those ridiculous light-up sneakers; it made me smile. An elderly black man laughed into his phone—a deep, resonant chuckle—and the sound made me happy. Vivien's face came back to me—her absorbed gaze the day before as we ate Mexican food at a riverside restaurant and she tweaked her presentation on her laptop; her smile and laugh as I teased her for eating the exact same foods she'd eaten as a child; her open warmth as she responded to questions after her talk.

Somewhere inside my chest, that pride I felt began to grow from the guttering, nearly extinguished flame to a blazing sun. The more I walked, the more the images cascaded in my mind: the way Viv would wear big ol' sneakers and tube socks in fourth grade; how she'd bring a stuffed chameleon toy named "Curla" to school and climb a tree with her friends to eat lunch in the branches. The sound of her big bag of colored pencils being rustled up on a weekend morning while her mother and I lay in bed. The shock of the cold salty Pacific when we dove into the surf together when we camped on Catalina Island.

I kept walking, back and forth. The blasting loudspeakers announcing flights didn't even register to me. The sun was fully up, shining brightly. So was I.

We all have a body. No matter who you are, if you're reading this, you have a body. This is the most basic fact of our shared existence I can think of. The way our body interacts with the world around us, and the way we perceive that interaction, is at the core of how we think, and who we are.

Admittedly, the San Antonio airport is an inauspicious place for a revelation. Yet when I was walking the airport, I was able to feel life deeply in a way I couldn't when I was sitting. The shift in my emotions, my belief about myself, and my feelings about the world came about through the simple, everyday miracle of movement. Simply walking changed my mood, and thus changed the way I thought about the world. I didn't have to cognitively work through how I was feeling; it just happened.

By returning to my body through the simple act of walking, I changed the way I thought, my emotions, and my behavior. Engagement with the physical self in a direct way had a domino effect that changed the way I felt about myself and the world.

In fact, a more accurate way to say it is that *we are our body*. Our concept of self is embodied, wired into the very physical tissues. In walking back and forth through the airport, I was simply reminding myself of this basic reality. Prior to returning to the physical realization that I am a body—that the experience of "me" wasn't some airy notion suspended in the clouds, but rather was the very corporeal stuff of my physical form—I was experiencing a split. I was too much in my head, and my sense of self was dislocated. It was as though my conception of self—who I was, how I felt, and my interaction with the world around me—was a balloon floating far above me. My walking, my breathing, and my return to the physical self hauled that balloon down. Popped it, even, and got me back into my body. We don't all simply have a body, we are bodies. Our bodies are us.

When I was a middle school teacher in Los Angeles a few lifetimes ago, I got the crazy idea to teach Shakespeare's *Hamlet*. I thought one of the ways I could really bring the play to life was to have the students act it out on camera, then we'd edit together a movie version. I have never taken a class in Shakespeare, nor would I consider myself particularly adept at parsing his linguistic gymnastics. But that play has always stuck with me, and a number of Hamlet's lines were burned into my memory.

One of them is, "The time is out of joint." Hamlet is referring to the way his sanity is suffering; after all, he's conversing with the ghost of his father and struggling with severe melancholy. Prior to reentering the present moment through the simple physical movement of my body in San Antonio, my time was out of joint, too. I was worried about the future, regretting the past. Replaying old arguments, imagining new ones. Building elaborate structures of anxiety, telling myself stories about how terrible everything was. I was not in the present; I was fragmented, existing as a shade (like Hamlet's father) and haunting my own past and future. All I needed were some chains draped around me and I'd have been my very own Dickensian ghost. Because I was out of touch with my body, I was also out of touch with the physicality of compassion—rather than seeing the shared humanity in those around me, my thoughts were small and petty.

But in the act of movement I found a salve. The worries and pressures I felt slowly evaporated. They weren't real, not really. I'd made them up. Yes, the world still suffered. Yes, life wasn't perfect. But I had regained myself in the present moment. It was all I needed to do to right the ship and set sail again, captain of my own destiny. Even if that destiny was only an egg sandwich from Au Bon Pain.

There's a name for the way in which our sense of self exists within our body. It's called *embodied cognition*. It's a wide field of study, and includes the effects of mindfulness meditation, the embodied nature of decision-making, how intuition works, and ideas about free will—the list goes on and on. There are quite a few places in education where a broad understanding of embodied cognition may be helpful.

There's a library full of books that provide additional practical advice on how bodies matter in school: Brad Johnson's *Learning on Your Feet: Incorporating Physical Activity into the K–8 Classroom*, *The Kinesthetic Classroom* by Traci Lengel and Mike Kuczala, and *Teaching through Movement* by Stacey Shoecraft. And, of course, the work of educational philosopher John Dewey. This book, however, is an attempt at exploring how a very basic reality (*we are our bodies*) affects learning—we'll try to get into the physical mechanisms that underpin much of our understanding about how bodies incorporate learning.

You may be asking, well, who cares? Yeah, we've all got a body. What does that have to do with school or teaching or how we remember information? Those are all questions I'll take my best shot at answering. But it's also deeper than that. Embodied cognition can help us understand how we even have a concept of "self" in the most profound sense. You know you exist, and have feelings and thoughts and emotions, but who is the "you" in that sentence? This book will try to convince the reader that the "you" is your body—and that your sense of your individuality is bound up in muscle and bone.

ONE
Bodies of Knowledge

How, exactly, do humans learn?

For much of Western history since the ancient Greeks, one of the central ideas underpinning much of philosophy has been that there is a separation between mind and body—that reason and rational thinking exist as abstract universals. This faulty dualistic view is pinned on Descartes and rightly so—it was the Cartesian duality of mind and body that fueled many of the ideas of the Enlightenment. It's only been over the past century or so that this Cartesian worldview of the disembodied mind has been largely overturned. Especially in the past few decades, a much more complicated and interesting picture of how the mind works in conjunction with the body has arisen through empirical findings in cognitive science.

This new way of looking at human consciousness and behavior and the principles that unify our mind and body is described as *embodied cognition*.

Put simply, *embodied cognition* is the conglomeration of a number of fields of study that posit that thought—including abstract ideas, reason, and rational analysis—are not disembodied universals but rather deeply embedded in our very flesh and bone. Embodied cognition research supports the idea that feelings and emotions are not obstacles to the process of rational thought, but part of it, inextricably intertwined.[1]

In a way, this can feel self-evident. For example, people will often claim to make decisions by "going with their gut." This embodied metaphor is a way of obliquely addressing the body's role in making complex choices. And the reality is that all decisions are "gut" decisions—they are formed, informed, and carried out by the body and mind in tandem. Particularly in education, decision-making of this nature becomes the daily act of learning. By teasing out the ramifications of embodied cognition, it becomes apparent that the way education is structured and the way teaching and learning are defined and enacted, are deeply affected by this emerging science.

It may be helpful in various ways to outline a practical approach to thinking about the relationship between our embodied experience and the inaccurate mind/body dualistic worldview. Take education, for example. What schools do on a daily basis, how educators interact with students and each other, the very buildings and spaces they teach in—all of these components of education can be addressed through recent findings in cognitive science and evaluated in terms of their relationship to embodied cognition.[2]

However, to unravel the tangled story of why education is structured the way it is, why teachers teach the way they do, and what the conception of knowledge is and how that affects the ability to learn, it is beneficial to look at the historical and conceptual underpinnings that frame those daily interactions. And it's within this intersection between what educators and students *do* in the act of learning and the ideological history of education that the powerful undercurrent of the ol' mind/body split can be found. At this juncture, the real work of bringing the hypothetical implications of an educational approach rooted in embodied cognition to the fore can begin.

At this point, it may be helpful to clarify what is meant by *embodiment*. In terms of learning and education, when the terms *embodied classroom* or *embodied lesson* are used, what is meant is that the *physical sensations of the body are taken into account and used as a meaningful method of learning*. Or, put another way, the sensorimotor perception of subjective experience is embraced as the means by which humans perceive, interpret, and interact with their environment.

Although embodied cognition is a relatively concise concept—thinking occurs in the body as well as the brain, and "mind" is really just a word we use to describe the interactions between body, brain, and environment—it is also an incredibly diverse chunk of research with many rich veins that bear on education and how people learn. As a way of offering a few preliminary examples of the sorts of ways embodied cognition could affect schools, a review of a few bits of research that potentially could have direct, practical influence on the way educators think about schools, teaching, and learning may be helpful.

WHERE YOU ARE: PROPRIOCEPTION

Most people don't have to think about where their bodies are. Humans just know. For example, when an individual is sitting and reading a book, legs tucked or splayed, book held or balanced on the midsection, that individual *knows* where their body parts are. One doesn't have to lift one's eyes from the page; humans have a sense known as *proprioception* that informs them of the location of their various body parts and of the position of their body in space. Even if a reader closes the eyes, he or she still knows it.

Humans know where they are in space, and where their limbs are in relation to one another. A person can scratch a cheek with eyes closed without fear of accidentally scratching an armpit instead. This seems simple to the point of inanity—of course humans know where their limbs are. However, there are rare cases of people losing their sense of proprioception—losing the ability to know their orientation in space and the placement of their body in relation to itself. But for most individuals, it's such "second nature" that it's a given.

Almost all people have this ability to perceive their body in space, but how do they do it? It's not conscious, or something one thinks about. It's not abstract, rational, or reasoned thought. People just *know*. This is one single aspect of embodied cognition—this ability is nothing short of amazing and is not separate from higher levels of thought or executive functions like decision-making, but actually part of those cognitive processes.

The only time there is even a taste of how crucial the sense of proprioception can be is when limbs fall asleep, not just a tingly foot at the movies, but those times when an arm gets slept on funny and completely loses feeling, much to the waking sleeper's dismay. The sleeper awakes to experience the limp, sensationless arm until the blood flows again and the arm electrifies itself back to life. That bizarre moment where one can't feel the arm is just the tiniest taste of what a full, systemic loss of proprioception might entail and strongly suggests how vital this sense is.

HOW YOU'RE FEELING: INTEROCEPTION

Whereas proprioception allows humans to understand the body's orientation in the environment, *interoception* is the sense of what's going on inside the body. When people ask a sick person, "How are you feeling?" they're calling, in part, on the individual's ability to express interoceptive capacity.

An excellent example of how this interoceptive sense can be assessed and examined is through an exercise known in mindfulness meditation as a "body scan." A body scan prompts individuals to "check in" with various parts of their bodies and see how they're feeling. Twisted? Sore? Out of joint, heavy, or imbalanced? Bilious? Achy? Interoception is the body's ability, through a complicated communication between multiple players such as the nervous, digestive, vestibular, and haptic systems, to assess how balanced the internal processes are.

Temperature and heart rate are good examples. The body is constantly seeking a state of *homeostasis*, wherein all operating systems are running smoothly. Of the many movement-oriented traditions that engage the interoceptive (and proprioceptive) senses, tai chi and yoga are excellent examples. Again, here is something that, on the surface, seems bulletproof simple. Of course humans know when internal viscera are out of whack—they feel crappy. But how do individuals "know" this? It isn't intellectual knowing or rational understanding. People "know" it in an embodied way. A human's interlocking systems know it together—the brain is just one of many switchboards, a communicative hub among many.

When individuals begin to develop our interoceptive abilities through practice in one of the movement-oriented traditions mentioned previously, one quickly realizes that the amount of sleep, blood sugar, interactions with other people in the recent past, the time of day, the season, and whether folks are indoors or outdoors (among many other factors) deeply affect the results of interoceptive scrutiny.

DANCE DANCE REVOLUTION: CORPOROCEPTION

Corporoception is a helpful term to describe a specific kind of proprioception—the embodied understanding of a body's relationship to other bodies. Although *corporoception* is a new term developed in the course of this research, it is empirically testable. It's the ability to perceive the configuration of the body relative to the orientation of other bodies. This is where fascinating research in proxemics comes into play.

In part, how close and in what arrangement bodies are placed with others depends on cultural norms and traditions but also on corporoception: humans' innate sense of other bodies. The most obvious example is dancers who are so "in tune" with one another that their moves seem harmonized beyond human conception.

This sense of bodily awareness is beautifully illustrated in the relationship between George Balanchine and Suzanne Farrell. Balanchine was the founder and director of the New York City Ballet and his work and vision dominated American ballet for much of the mid-twentieth century. Suzanne Farrell (born Suzi Flicker) was his principal dancer for many years. The work that they achieved together has gone down in dance history as one of the most incredible artistic pairings ever achieved, regardless of genre. In his book on creativity, *Powers of Two*, author Joshua Wolf Shenk notes that the two achieved a kind of "confluence."[3] This kinesthetic understanding was so profound that Balanchine could offer Farrell the most intrepid, wild challenges; she often met them and exceeded them. In one instance, Balanchine prompted Farrell to attempt a spin from an unusually wide fourth position (for the non–ballet dancers out there, that's like trying to jump and spin with your legs crossed and spread wide, destabilizing balance). She does it, and later called it a "turn of faith."

What's going on with Balanchine and Farrell? Shenk notes that creative pairings often have a similar trajectory in their bonding: "presence leads to confidence, and confidence settles into trust, trust elevates into faith."[4] Note that to have that level of corporoceptive understanding, the first step is "presence"—physical, actual presence, bodies in a room. In fact, that's what makes dance so exhilarating to watch: it's live, and there is something about being in the presence of bodies that are so finely tuned to one another that is breathtaking, similar to the experience of being present for performances by incredible musicians like Yo-Yo Ma or Jimi Hendrix or hearing brilliant orators like Dr. Martin Luther King Jr.

Although ballet is one example, community dance, or improvised, discursive dance, is perhaps an even better illustration of how bodies can read intent, mode, and feeling from each other, and the role a deep intuition can play in dance. African American dance traditions, in particular, showcase this incredible kinesthetic give and take during improvised, communal dances. A large number of helpful examples are on YouTube, where viewers can watch old *Soul Train* dance lines.[5] Here, we have a large number of dancers who are dancing communally, keeping the beat and rhythm together, as well as improvising dance routines as they take their turn moving down the line. The back and forth, improvisational nature of such scenes can be found in multiple forms throughout African American dance communities. Similar to the call-and-response style of sermon traditionally given by black Christian pastors, this give and take represents a complex corporoceptive understanding—a deeply empathetic relationship that draws on the dancers' intuition and "read" of other bodies. For a more recent example (for those millennials who just went blank at the mention of *Soul Train*), the video for "Watch Me," a song by rapper Silentó, is a perfect example of communal dance, where iterations of the same steps are tweaked and improvised and tossed back and forth in an ever-changing conversation among bodies. In an essay on embodiment titled "The Black Beat Made Visible," Thomas DeFrantz rhetorically asks, "How are similar dances known so thoroughly by dancers who have never met; who have had no cause to collaborate?"[6]

The rhizomatic exchange of kinesthetic colloquialisms within the black community is perhaps the richest example of this corporoceptive ability—an embodied cognitive element that allows for dancers to "know" each other when they've in fact never met. Rather than more traditional hierarchical forms of knowledge transfer, the corporoceptive understanding of bodies pops up here and there, deep roots spreading the moves and dances in the same way that bamboo forests are not individual trees but rather a vast underground rhizomatic root system giving rise to each individual shoot. And although the internet certainly helps diffuse new dances quickly, it is still very much an embodied transmission of knowledge, a physical exchange of ideas and expression.

This facet of embodied cognition becomes particularly important to consider when hypothesizing what the effects might be of an entire generation of children who grow up playing video games rather than physically wrestling and tumbling with other bodies. What level of consciousness—or, for that matter, kinesthetic intuition—is lost when the primary vehicle for interpersonal relationships is not the corporoceptive interchange of embodied missives but the flat world of the digital screen?

But this idea is even more crucial to understand when we consider that humans are social animals. When in contact with other humans, bodies respond on multiple molecular, chemical, emotional, physical, and neurological levels. When parents snuggle with their children on the couch to read a book together, or when couples intertwine their feet under the blanket on a cold night, the body produces oxytocin, a hormone that responds to physical touch and encourages those social bonds. In this way, the sense of embodiment is *dependent on other bodies.*

Individuals who are in solitary confinement for long periods suffer from various health issues; infants and toddlers who aren't held and physically touched have developmental issues, as the horrifying stories that came out of orphanages in Romania during Ceaușescu's dictatorship attest.[7] Children who were denied vital bodily contact in understaffed orphanages in the 1980s had cognitive delays, behavioral issues, and health problems. Many of the sad stories connected to Ceaușescu's orphanages can be directly related to the lack of touch those kids experienced.

Corporoception, the sense of other bodies, isn't just the senses picking up on other bodies nearby. It is the internal processes of various systems responding to that very closeness—a neurochemical tractor beam, so to speak. Humans need to be around people; they need to be touched. Those who doubt the power of close touch and call into question the way that two bodies can act on each other would benefit from watching some primate grooming videos on YouTube or reading Frans de Waal's recent book *Mama's Last Hug,*[8] the title of which deals explicitly with primates' need for close, affectionate contact. It's always striking to see the complete and utter blissed-out relaxation of a primate being groomed—and humans are not different from those monkeys as far as corporoceptive needs are concerned.

THE EMBODIED SELF

It is not that educators need to develop both the body *and* the mind through curricula and pedagogy. Teachers need to do both because they are the same thing, interwoven parts of one system. There's nothing particularly earth-shattering about this—Waldorf and Montessori schools, to name just two, have valued "hands-on learning" for decades, and one of the dons of educational philosophy, John Dewey, was a proponent of just this kind of experiential learning.[9] What has changed is the contemporary understanding of the empirical evidence behind the notion of learning through the body.

Research from Andrew Wilson and Sabrina Golunka from Leeds Metropolitan University in the United Kingdom puts it thus: "Our behavior emerges from the real-time interplay of task-specific resources distributed across the brain, body, and environment, coupled together via our perceptual systems."[10] This suggests that, although there is an assumption that something like "decision-making" is a purely mind-related concept, in actuality making judgments about complicated intellectual and moral problems are embodied, and the interconnected systems of the body replaces the need for "complex internal mental representations."[11]

How do humans have a sense of Self—an understanding of individuality in the world? This is a complex philosophical question that has left many an undergrad scurrying for the nearest distracting game of Halo Infinite. Descartes answered by suggesting that, *because he could think, he existed*. Other essayists, psychologists, and scientists have tried to define this elusive idea of "selfhood" and "the mind" and discover its roots. But in exploring the idea of Self from an embodied-cognition perspective, the picture snaps into focus. Olivier Gapenne from the Université de Technologie de Compiegne in France writes that proprioception, the felt sense of the orientation of the body in relation to the environment, is a key to understanding how we conceive of "Self":

> When one wishes to account for the constitution of the distinction between the self and the world there is a necessity for the acting agent to make a distinction between two sources of variation in the sensory signals that affect it: those that are related to its own activity, and those that arise from the environment.[12]

This quote suggests that the concept of Self, the very way individuals identify as themselves, and the reason we assume any personal meaning at all, is due to the body's orientation in space and the fact that the body *knows* its orientation in space.

Another way to think about this is to consider that cognition isn't something occurring in some imaginary space that doesn't exist. "She's in her head" is a phrase used to describe people experiencing deep thought or reverie, but it's inaccurate in the extreme. Not only are thoughts affected by embodied experiences, they may be defined by experiences as well.

AN EMBODIED EDUCATION

How did embodied cognition come about? Is it a recent evolutionary development or something hardwired? A researcher named Aaron Stutz believes that the development of a cognition that is embodied and a human's ability to think abstractly were linked through the same evolutionary circumstances:

> Human capacities for symbolic mental representation, symbolic communication, and social cooperation emerged over the past ca. 5–7 million years through dynamic co-evolution with embodied cognition and environmental interactions.[13]

The effect these new developments within cognitive science could have on education is profound. For the past few hundred years, the very idea of learning has been bound up in the Cartesian duality of mind/body. Learning has been seen as an activity of the mind, as the strengthening of some abstract understanding of reason and a purely rational process. But through embodied cognition, this concept gets turned on its head. Rather than the false dichotomy of pure reason versus the more tangled, corporeal stuff of the body, cognitive science is forcing educators to rethink the way teaching works: through the body, not in spite of it.

The way embodied cognition interacts with a more traditional concept of education and learning can be roughly broken into a handful of categories: movement, feelings, thinking, meaning, and reason. What follows is a brief introduction to each of these concepts with the aim of looking at how these aspects play a role in both traditional educational practices as well as learning based in embodied cognition.

I LIKE TO MOVE IT, MOVE IT

Guy Claxton has written widely about education. In a recent book titled *Intelligence in the Flesh*, Claxton distills the way embodied cognition influences education in the following statement: "[W]e are fundamentally built for action."[14] We can infer that the traditional school that keeps students sedentary, indoors, and inhibiting movement may be diametrically opposed to the way cognition and learning work.

The cultural divide between the way in which physically complex and sophisticated abilities and intellectually complicated tasks are viewed in relation to one another is vast. The idea of "the dumb jock" is pervasive in entertainment and education. But the reality is far more nuanced. It is not that skateboarding kids who can bust a switch kick flip down a flight of ten stairs are less intelligent than the studious AP calculus pupil; it's simply that throughout history humans have engineered a society that favors nonphysical expressions of cognition—something that Claxton calls "the hegemony of intellect."[15]

Movement is so central to learning and cognition that the two are really the same thing; it's not that learning has the opportunity to occur through movement, it's the idea that learning *is* movement. As Maxine Sheets-Johnson puts it, "[W]e literally discover ourselves in movement."[16] In fact, not only can humans come to know themselves, or even our symbolic "Self" through movement, but moving the body is inextricably tied to understanding complex ideas. Philosopher Mark Johnson puts it this way: "[W]hat we call abstract concepts are defined by systematic mappings from body-based, sensorimotor source domains onto abstract target domains."[17] Or, put more simply, complex ideas can be more thoroughly understood through movement.

REASONABLE PEOPLE CAN DISAGREE

Logic and reason are the twin pillars that hold up much of Western philosophy. From the ancient world of the Greeks and Romans through the Enlightenment, reason is often thought of as humanity's finest achievement, the very thing that *makes* people human beings. Reason was predominantly seen in the abstract, meaning it was viewed as a set of universal laws, governing some clockwork-like logic upon which the axis of existence turned. Reason was seen much the way God is often viewed in the Judeo-Christian world—as the structural organizing principle of the universe.

It turns out that cognitive science has been able to take reason out of the airy halls of abstract philosophy and embody it in human beings' lumpy, hirsute, imperfect forms. Mark Johnson—mentioned earlier—wrote a book with another philosopher named George Lakoff called *Philosophy in the Flesh* providing convincing evidence that reason is bound up in the body, from the roots of hair to the tips of the toes. The authors claim that reason "is shaped crucially by the peculiarities of our human bodies" as well as the "specifics of our everyday functioning in the world."[18]

Reason doesn't exist on some mathematical, astral plane: it's in the very bones. Individuals reason through and with their bodies, not in spite of them. This is a key shift, particularly for teachers. In the most basic sense, if educators are not engaging students physically, in their own bodies, and stretching and strengthening and engaging them with the world around them, they are robbing them of the chance to fully use and explore the capacities of their embodied reasoning. They are limiting the students' learning.

THE EMBODIED CLASSROOM

Rather than just theorize, it may be helpful to jump into how embodied cognition might look if it was incorporated into the daily life of an elementary school student. But to do so, it's essential to first comprehend the way embodiment informs the way human beings think.

Johnson and Lakoff argue that all reason and abstract thinking are metaphorical, and that most metaphors are embodied, meaning based in physical form and that form's relation to the surrounding world.[19]

Think, for a moment, of the phrasing that started this particular section. The first sentences stated that the reader could "jump into" ideas about embodied cognition. The understanding and conception of how individuals begin to learn is best described in an embodied metaphor—it is, after all, bodies that jump, and real physical selves that "get into" things, places, and ideas. This is also about movement. It's not by accident that the expression states that the reader "jump" into ideas rather than "lay on top of" or "sit" on an idea. Learning is defined with bodily and exploratory metaphor. The way progress is perceived in learning (even progress can be seen as an embodied metaphor) is as movement toward something.

Back to the classroom, where students are studying math. The Romans took the most literal and direct route possible in mathematics by breaking all numbers down into blocks of ten—something that still occurs in classrooms today. The speedometer of a car, for instance, probably reads 10, 20, 30, 40 miles per hour, and so on. The reason is simple: most humans have ten fingers.

Even in high school, when students start studying higher mathematics like algebra, embodiment is woven through everything. The Cartesian grid of the X and Y axes is a directional equation; add a Z axis and the students are now talking about and working with three-dimensional space, or proprioception. And even though this level of computing is within the realm of higher math, the route to comprehension is still the body.

The teacher in an embodied classroom begins the day with Tai Chi. Students stand and the teacher leads them through the fluid motions, working with them to focus and hone their proprioceptive and interoceptive senses and fine-tune their vestibular systems. The classroom is filled with the teacher's voice, and the students will, at first, giggle and laugh and struggle and resist the demands of graceful, balanced movement. But that's the way they respond to any new learning construct.

Asian cultures seem to understand the inextricable nature of embodied cognition in some intrinsic way that the Western world—dominated by ideas from the Enlightenment—does not. Aikido, Tai Chi, Zen meditation, kung fu, yoga—all of these traditions have a similar central tenet of being in the body, through movement or breathing, and locating an actual physical point or location within the body: one's *chi* or *ki*, where "centeredness" exists. One wonders what the benefit would be for those students who seem like living pinballs in finding that calm, solid, strong place within themselves from which to act, rather than always reacting to their environment.

The students take forever—maybe days, maybe weeks—to finally get into the swing of things with the Tai Chi or yoga or whatever movement-oriented embodied experience the teacher has introduced. But finally they do, and focusing on being within their own bodies, honing their sense of embodiment, becomes part of the routine. The usual rationale for such exercises or activities is simply health—movement is good for circulation. It gets the "yayas" out. But it's actually functioning on a much deeper neurophysical level.

Because functional thought is largely metaphoric, and metaphors are mostly embodied, the only way to really understand conceptual ideas is to be *in* the body—to fully inhabit one's physical self.

It's not only that Tai Chi has health benefits and a healthy person can learn better—although that's true and convincing enough as a reason to pursue an embodied cognition approach to learning—but rather that Tai Chi incorporates fundamental concepts of proprioception and interoception that, in turn, allow for a deeper, more "real" comprehension of abstract thought.

The students finish their Tai Chi and move on to language arts, fresh from their martial art. Now, they are innately in tune with their sensorimotor experience. Sensorimotor experiences are simply the lived relationship between the somatic senses and the body's movement through the world. In the ensuing lesson, the students' achievement is higher because they are now attuned to their bodies, which is the way they process cognitively. The body and brain are both engaged in learning. In fact, as the teacher introduces complex sentences, or punctuation, or how to create a narrative, the very work of the students in processing language is done from a sensorimotor perspective.

In this scenario, adjectives are the lesson of the day. Now, if a student has never heard or spelled the word "squishy," the bodily experience of holding something squishy while learning the word has been shown to increase retention of the word in the students' vocabulary as well as the likelihood that they'll use the word in their own writing.[20] Or, in another scenario that has been proven effective, children who learn a story and then get to act it out retain and comprehend the information more than those who simply read and reread the story.[21] Embodying principles—whether it's a simple children's story or the negotiations at the United Nations—allows students to embody their learning; this embodiment in turn fosters learning.

Embodied cognition can manifest itself in multiple ways. For instance, research has shown that children who stick their tongue out during the processing of affective concepts were able to process those concepts more quickly. How humans engage with their bodies—*down to their very tongues*—matters. By and large, these lessons have been ignored by schools, and even the medical community has at times disregarded the importance of embodied experience and movement—to the point at which those phenomena are pathologized. Many students fidget and squirm—an antsy jostling of the foot, leg, or body—in a way that seems almost unconscious. Perhaps this is simply students trying to filter learning through their very bodies—trying to rhythmically bounce knowledge through their very being.

Left to their own devices, children generally are naturally embodied; that is, they are in tune and acting in accordance to the dictates of their physical reality. Large swaths of unstructured time outdoors is possibly the most efficient, expedient, and practical way to provide an appropriate setting to engage students in embodied activities. Constructed, institutional spaces are usually severely limiting in terms of offering opportunities for movement that fully utilizes physical cognition. Children who play in the woods, fields, and vacant lots, however, are constantly engaged in a direct way with the world around them. Consider the decisions made by a kid building a dam of river rocks in a stream: lift this rock this way, place more stones here to shore up this section, step here where it's less slippery; these are the same structural processes that govern more recognizably cognitive decisions. Which books to use to research a history project, which quote to use, the right word choice in an essay—all these decisions operate on the same pathways built into the body that are used in physically negotiating the environment. Thus, giving students gobs of time outdoors to explore, adventure, and play is vital to developing decision-making capacities.

Taking students outdoors can often seem like "a waste of time" if looked at through the lens of contemporary educational policy. It can seem at times that schools suffer from a nineteenth-century hangover, bent on production and industry and a general busyness. However, when using a different lens while watching what seems like idle playtime, it becomes clear that a group of third graders, or ninth graders, building forts or climbing trees or playing in the waves are bodily engaged in their environment in a manner that speaks directly to humans' inherent bodily systems that govern behavior.

What looks like aimless playing and goofing off may in fact be the most crucial activity a child can engage in.

NOTES

1. Lawrence A. Shapiro, *The Routledge Handbook of Embodied Cognition* (New York: Routledge, Taylor & Francis Group, 2017).
2. Andrew D. Wilson and Sabrina Golonka, "Embodied Cognition Is Not What You Think It Is." *Frontiers in Psychology*, vol. 4 (Feb. 12, 2013).
3. Joshua Wolf Shenk, *Powers of Two: Finding the Essence of Innovation in Creative Pairs* (London: John Murray, 2015).
4. Shenk, *Powers of Two*, 35.
5. https://www.youtube.com/watch?v=4dIhCdIK4Fc
6. Thomas DeFrantz, "The Black Beat Made Visible."
7. Jane Perlez. "Romanian 'Orphans': Prisoners of Their Cribs." *New York Times* (Mar. 25, 1996).
8. Frans de Waal, *Mama's Last Hug: Animal Emotions and What They Tell Us about Ourselves* (New York: W. W. Norton, 2019).
9. John Dewey, *Democracy and Education* (Radford, VA: SMK Books, 2018).

10. Andrew Wilson and Sabrina Golunka, "Embodied Cognition Is Not What You Think It Is," *Frontiers in Psychology*, vol. 4 (Feb. 12, 2013).

11. Wilson and Golunka, "Embodied Cognition Is Not What You Think It Is."

12. Olivier Gapeene, "The Co-Constitution of the Self and the World: Action and Proprioceptive Coupling." *Frontiers in Psychology* (June 12, 2014).

13. Aaron Stutz, "Embodied Niche Construction in the Hominin Lineage: Semiotic Structure and Sustained Attention in Human Embodied Cognition." *Frontiers in Psychology* (Aug. 1, 2014).

14. Guy Claxton, *Intelligence in the Flesh* (New Haven, CT: Yale University Press, 2016).

15. Claxton, *Intelligence in the Flesh*, 9.

16. Maxine Sheets-Johnstone, *The Primacy of Movement* (Amsterdam/Philadelphia: John Benjamins, 1999).

17. Johnson, *The Meaning of the Body*, 177.

18. George Lakoff and Mark Johnson, *Philosophy in the Flesh: the Embodied Mind and Its Challenge to Western Thought* (New York: Basic Books, 1999), 4.

19. Lakoff and Johnson, *Philosophy in the Flesh*.

20. "Touch Can Produce Detailed, Lasting Memories." *Neuroscience News* (Nov. 27, 2018).

21. Arthur M. Glenberg, et al. "Use-Induced Motor Plasticity Affects the Processing of Abstract and Concrete Language." *Current Biology* vol. 18, no. 7 (2008).

TWO
A Brief History of a Good Idea at the Time

Cartesian duality would have it that to have knowledge—to possess facts and information, to understand concepts, and to be able to articulate arguments—is a construct of reason and rationality. However, even only a bit of travel along that line of argument quickly runs into trouble because abstract, purely theoretical knowledge only gets so far. It's incomplete.

For example, take an account of an early European "explorer" who recorded a conversation with a member of one of the Inuit tribes of the northwest latitudes in Greenland. This particular Inuit had never seen trees, as they don't grow in those frozen Arctic lands. The explorer related the futility of trying to explain what a tree is, how it looks, to someone who'd never seen one. How does one get across the idea of "treeness" to an individual who's never experienced it? The Inuit could not believe the outlandishness of a massive, one-hundred-foot-tall organism that reached deep into the earth with its many-limbed arms as well as into the sky, and that shed its leafy verdure every year and became a skeleton, only to burst back into life later on. (Of course, this story is probably apocryphal; it is easy to imagine the Inuit was pulling the wool over the gullible colonizers' eyes, and later on had a good laugh with the family back at the igloo over his trick). But the story helps illustrate a point about bodily versus abstract knowledge, pure cognitive understanding.

Here's another example, clunky perhaps, but offered to drive the point home. Imagine an individual who has never seen a cow. Someone tries to describe this big, half-ton animal; it's got a fleshy bag of milk on its underside, horns, and a swishing tail. It has cloven hooves and chews cud. There are all sorts of things that could be used to describe a cow to an individual who hasn't seen one. Perhaps, if they'd seen a horse, they'd have a better idea, or better yet a yak or buffalo. But nothing that could be offered in terms of knowledge and information would compare to the sensory experience of seeing, touching, smelling, and hearing a cow: the low, cello-like moan of their lowing; the acrid reek of cow urine and the peaty musk of manure; their warmth and size; their large eyes, softly lashed. To really consider a cow, up close, and run a hand along its head and neck, or feel that massive, sandpapery tongue, is to think to oneself that cow-ness is just not a concept one can think about in the abstract. One has to experience it.

In his rich and fascinating 2017 book, *The Evolution of Beauty*, ornithologist Richard Prum sums up this dichotomy between knowing on a purely mental level and experiencing something with the sense, with the body.[1] For Prum, a rabid, lifelong birder, to know a warbler—to know its size, colors, taxonomy, and habitat—is one thing. To see it on the wing, however, is something else entirely. In fact, birders in general are obsessive about seeing the actual animals rather than just cataloging them. At the core of their love of birds is the experience of heading out, binoculars and field guide in hand, to try to see feathered specimens in the flesh. Darwin was a birder, to a certain extent. So was his contemporary and competitor for the discovery of the theory of natural selection, Alfred Russel Wallace.

Prum notes that other languages have separate verbs for just "knowing" versus "knowing through experience."[2] In Spanish, *saber* is to know a fact; *conocer* is to know something through direct experience. "It's about accumulating knowledge about the natural world through your own personal experience," Prum wrote. "Knowledge without experience," he adds, "is never enough."[3]

MIND/BODY DUALISM

Although Descartes's theory of the separation of mind and body gets a fair amount of press as an explanation for why Western institutions think about cognition as they do, it seems appropriate to dig into what, exactly, Descartes was trying to say. What was his motivation in separating consciousness from the physical form? It's of particular importance for the purposes of this book as his ideas, and their antecedents, inform to a considerable degree how educators talk about learning and education to this day.

Rene Descartes was a French philosopher who spent much of his adult life in what was the Dutch Republic of the early seventeenth century. He is most famous for the quip, "I think, therefore I am," from his book, *Principles of Philosophy*, from 1644.[4] He was committed to separating the mind and the body for various reasons—some easier to understand than others.

Descartes believed that the mind is completely different than the body. By separating the two in terms of their fundamental nature, Descartes argues that they can then exist in isolation. In other words, if their core elements are completely different, they must be able to exist on their own, because they don't depend on one another in any way. It can start to make one's head spin. But in the most basic design, Descartes believed that there was a difference between substances. By *substances* he meant some sort of fundamental thing, separate from all other things and irreducible. Then, there is the mode of that thing—basically whatever it does in the world. For example, a stone is a substance in and of itself, and can exist all alone without anything else, and heaviness is its mode. He applies this logic to the mind and body, suggesting that they are both substances and therefore can exist in isolation, apart from one another.

Many within the church viewed Descartes with suspicion. He had been friends with Galileo, and had seen what happened to him when he posited that the earth revolved around the sun instead of vice versa, so Descartes was pretty assiduous about including God in his calculations about how the universe worked. However, the driving point behind Descartes's famous "I think, therefore I am" statement was doubt. The only thing that we can be sure of, according to Descartes, is that we are thinking thoughts. Everything else—the world around us, our own bodies, the arbitrary laws of the universe that we devise—could all be illusions, the sneaky trick of an evil genius. Essentially, Descartes believed we could be in the Matrix, like Keanu Reeves's character in the eponymous movie, and the permanence and reality of everything we see can be doubted—except, of course, thinking, hence Descartes's famous line. The religious verifications came after making peace with the fact that we can doubt the reality of just about anything.

But Descartes was religious, and one of the benefits of this mind/body separation was that it supported the idea of an immortal soul belonging to God. For Descartes, mind and soul were the same thing; therefore, if he could prove that the mind/soul was somehow separate from the body, that would support the idea of life after death within the Christian tradition. In addition, Descartes was also part of the growing Enlightenment ideas of the time. He was at heart a mathematician. It's likely students probably run into him at some point in their early schooling during algebra when they plot lines on a Cartesian grid. So they can thank Descartes for that. But whether students enjoyed the X and Y axis or not, the fact is that Descartes had a pretty snarled problem to untangle: trying to validate both the idea of an eternal soul as well as the mechanistic physics of the world. Supporting the church and science at the same time has never been easy; Descartes's separation of mind and body was his attempt at doing so.

The Enlightenment really came into full bloom one hundred years after Descartes. But many of the seeds of the Enlightenment were laid during his time. In addition, the structure of school, how educators in the West think about education and knowledge, and the definitions of learning, progress, and intellectual advancement are deeply rooted in ideas that have their origin during the Enlightenment. Reason, the scientific method, rationality, and intellectual achievement all stem in some ways from advances made during the Enlightenment. Those structures, such as schools and the way educators and teachers think about consciousness and the mind, were based on the early work of Descartes. To this day, students sit at desks, immobile, to learn.

SENSUOUS BODIES

If humans don't just *have* bodies, but *are* their bodies, then it is the body's relationship with the world around us that defines who humans are. The air, the earth, nonhuman animals, smells, sounds, light—all play a vital role in shaping the experience of the world, and thus the sense of Self. David Abram's book, *The Spell of the Sensuous*, is a beautiful meditation on this idea. Abram writes: "Our direct experience is necessarily subjective, necessarily relative to our own position or place in the midst of things, to our particular desires, tastes, and concerns."[5] It is this subjectivity that created a bit of a conundrum for Enlightenment-era science and all the theories spawned during that time.

The desire for objectivity goes hand in hand with the scientific method. But if the experience of the world is the primary interface with a sense of "self," then all inferences of the world around us are skewed by human's subjective natures. This was the sort of tail-chasing that Descartes and others tried hard to avoid.

Immanuel Kant picked up some of Descartes's work, tweaking it so that some of the shortcomings were addressed.[6] During the late eighteenth century in Germany, Kant focused on the phenomenal world. By *phenomenal*, he meant that the nature of reality is our experiences. The appearance of the world, the way we perceive it, is the phenomenological view. But the true essence of things—reality, truth, call it what you will—can never be known. Kant called this *noumena*, which are basically the presumed actual elements of existence that we'll never really know about. This viewpoint was further refined by Hegel, yet another German (you'd think with all that great beer, Germans would be more relaxed about searching after the truth of existence) who worked up until the mid-nineteenth century. The point of all of this is that there was a movement in philosophy toward reincorporating our subjective experience of the world into reality. Rather than the cold, disembodied split of Descartes, philosophers were trying to account for our daily lived and perceived experience. This is crucial for understanding the way in which schools treat knowledge and learning. By and large, educational institutions have yet to incorporate all the ideas around truth and how humans think and perceive the world that's come after Descartes.

Edmund Husserl, a late nineteenth–early twentieth century German philosopher, called the subjective experience of the world around us *Lebenswelt*, which translates as the "life-world."[7] Husserl was nostalgic for "universal truth." In a way, he was trying to finish Descartes's work, looking for that one single truth from which the entire tapestry of existence can be woven. Like Kant before him, Husserl was a phenomenologist who tried to shore up human understanding of lived experience in the messy, nonrational world with a system of logic that would be universal. In a nutshell, Husserl believed that the *Lebenswelt* was the direct experience, the reality, of the world around us, which includes our past experiences, our memories, and the way in which we have, and do, perceive reality. It's life and world at once, in the present moment. Husserl amended Descartes's thinking by offering up the idea that, yes, we have thoughts, but they always have an object. "I love this tomato" or "I'm scared of this dog" have objects attached to the thought. Husserl attaches the idea of disembodied thought to the world of objects.

All this head spinning is why, when you look into an undergraduate philosophy course, you see lots of slack jaws, drool, and vacant expressions; it's enough to fry a person's brain. An example may help to understand the crucial step Husserl and others took to begin to incorporate the body into the way cognition and thinking are understood.

Two students are headed into a test. It's a big deal—the final exam. One student loves tests; she's the sort of type-A, organized person who can intuit what will be on the exam, studies hard, and always does well. Her experience of tests has been great, and she's often singled out and praised by teachers because of how well she does. The other student? Not so much. She flunked the midterm and feels overwhelmed when faced with organizing all the information needed. She can't memorize all that well. The feeling this second student has on entering the test room is sheer terror, fear that she'll let herself, the teacher, and her family down when she bombs the exam.

Husserl would say that the "natural standpoint" is that they are taking the same test. "Natural standpoint" is Husserl's way of suggesting that the objective world doesn't really exist. Sure, the test the two girls will take is the same, but their experience of it is markedly different. Because of the *Lebenswelt*, it's two different tests when viewed through the subjective lens of personal experience. Husserl sees the way humans subjectively perceive the world around them as a set of possibilities. It is not the same test for the two students. In fact, Husserl would argue, it is a very different test for each of them. And yet Husserl's view still retains the "truth" that only one test exists; this is his attempt at balancing the pure logic of Descartes with the messiness of phenomenological thinking introduced by Kant, Hegel, and others. This "intersubjectivity" of perception and experience lays the vital groundwork for an embodied view of cognition, because it offers up sensory perception of the world as a vital ingredient in the stew of how thought, learning, and experience are described from a philosophic point of view. To see the *Lebenswelt*, one tunes into the world, taking note of the way past experiences inform present perception.

This isn't to say that to experience the life-world, individuals need to sit in the lotus position on a mountaintop. Quite the contrary; it's there in everyday moments of tooth brushing, coffee making, and taking out the garbage. It's the sidewalk underfoot and the sun overhead, the world individuals depend on and know is there but never really think about. And yet it is this world and humans' unique and subjective position in it that is in constant interplay and flux with thoughts on life and death and everything in between. For much of Western history, thinkers have assumed that the life-world is apart from those magisterial ruminations, and yet they aren't. Logic and rationality and thought and learning don't exist in some pure vacuum as truth unsullied by the chaos of the world. Those ruminations come from the very subjective moments of individual experiences.

Abram's book riffs wonderfully on Husserl's theory: "If the world's experiences by humans are so diverse, how much more diverse, still, must be the life-worlds of other animals—of wolves, or owls, or a community of bees!"[8] This subjective experience of the life-world of nonhuman animals is a more radical notion still, diverging even further from Enlightenment ideas.

Descartes, who, like many of his contemporaries, saw animals as little more than machines, would've flipped out. All one needs to do is head outside with a dog for a walk on any given day and Abram's point snaps into focus. Any dog would step outside on a breezy spring day and be immersed in sensual stimuli. The wind blows and lifts the hairs on her shaggy coat, and she stands still and seems to be feeling that wind, absorbing some message about weather or direction or temperature.

One occasionally gets a sense of what animals may feel when a breeze delightfully runs up the neck, and the caressing breeze is experienced as this incredibly delightful tickle on the scalp. Dogs often stand at first once they've trotted outside, stretching out their necks and sniffing and snorting the air. From the human perspective, one wonders what it's like to absorb one's surroundings so completely through a sense of smell. A dog will stand stock-still, the world swirling around her, experiencing her own "life-world" that is different than that of a human's, the subjective experiences divergent, and yet with similarities, too.

Husserl's torch was picked up by the French philosopher Maurice Merleau-Ponty, who, during the twentieth century, solidly placed the body at the center of experience. Abram parses the dense texts of Merleau-Ponty and offers this succinct account of his thinking:

> If this body is my very presence in the world, if it is the body that alone enables me to enter into relations with other presences, if without these eyes, this voice, or these hands I would be unable to see, to taste, and to touch things, or to be touched by them—if without this body, in other words, there would be no possibility of experience—then the body itself is the true subject of experience. Merleau-Ponty begins, then, by identifying the subject—the experiencing "self"—with the bodily organism.[9]

Throughout children's experience in school, there is a steady devaluation of the bodily experience of Husserl's "life-world." As kids progress through elementary and middle school, they get less and less direct experience with the tactile world, and are more and more immersed in abstract reasoning and detached cognitive experiences. In fact, somewhat paradoxically, the very idea of bodies becomes icky and awkward for many kids to talk about. Perhaps a part of this is due to the collective Puritan hangover in America, but may also have to do with the reality that direct engagement in the "life-world" has taken a back seat to more cognitive mental reasoning processes. Even these processes, however, are embodied.

A fair question is, why? If all the tweedy professors at universities around the world have dug into the Kant and Hegel and Husserl and Merleau-Ponty, there must be some acknowledgment of the way in which the body and the perception of the world play central roles in how experience is transmuted into thinking, consciousness, and a sense of Self. Why do most institutions, from Shanghai to Chicago, default to a disembodied, nonsensory approach to education in which students are usually immobile, teachers drone on, screens flicker, and the body is forgotten?

Possibly, the industrial, transactional model of education is so powerful that for schools to break out of the conveyor-belt mode of instruction is impossible. The cheapest, most familiar, and streamlined way to teach thirty-five children about tectonic plates is to sit them down and explain it, show a video, and then test them on vocabulary: *mid-Atlantic ridge, lithosphere, mantle*. Disembodied intellect is favored over embodied understanding because it's easy.

But hope abounds. Spending time with elementary school teachers is often a healthy reminder of how important it is to bring some level of embodiment to learning. Many grade school teachers have kids acting out scenes from books, playing math games with blocks, building potato clocks and drawing pictures of their ideas. All too quickly, however, classrooms become more and more stagnant, stationary, and sedentary. Students are given less and less opportunity to use their physical, sensorimotor engagement with the world as the framework through which they learn and grow. Embodiment is the venue through which intellectual personal development progresses.

But without acknowledging the reality of this embodiment, the concrete realness of the world fades into the background, and, both literally and figuratively, humans "lose touch" with reality. Bodily reality is no longer the primary lens through which individuals experience the world; students without the opportunity to engage in the sensuous are left looking at their lives through the wrong end of a spyglass.

NOTES

1. Richard Prum, *The Evolution of Beauty: How Darwin's Forgotten Theory of Mate Choice Shapes the Animal World—and Us* (New York: Anchor Books, 2018).
2. Prum, *The Evolution of Beauty*, 5.
3. Prum, *The Evolution of Beauty*, 5.
4. René Descartes, *Principles of Philosophy*, Part 1, Article 7.
5. David Abram, *The Spell of the Sensuous* (New York: Vintage, 1996), 32.
6. Emmanuel Kant and Allen W. Wood, *Basic Writings of Kant* (New York: Penguin Random House/Modern Library, 2001).
7. Edmund Husserl and William P. Alston. *The Idea of Phenomenology* (Dordrecht, Netherlands: Kluwer Academic, 2010).
8. Abram, *The Spell of the Sensuous*, 42.
9. Abram, *The Spell of the Sensuous*, 42.

THREE
Evolutionary Baggage Claim

It can't be overstated that the current predicament of human life was shaped by millions of years of evolution. This is such a basic truth that it can be a little difficult to grab hold of. Like air, the reality of human adaptation over the course of the species' evolution is all around yet invisible most of the time. This reality is often taken for granted, and yet understanding the way in which humanity's history as a species has shaped bodies, minds, and behaviors is key in figuring out how and why humans learn, interact with one another, and experience the world.

Put another way, individual people are not just plunked into existence the moment they're born, as a tabula rasa. Rather, they come with a whole mess of systems and structures and instincts that were created and modified over millions of years, during which lives were radically different from what most people experience today. The only exceptions are those who recently narrowly escaped getting eaten by a saber-toothed tiger, headed back to the cave, and hacked out some delicious bone marrow with a stone for breakfast this morning. Most folks had coffee and a bagel.

This reality becomes particularly important when considering the way in which human bodies interact with the environment, download that experience neurologically, and then act accordingly. But already this linear equation is rife with problems: is it the body telling the brain what's going on, the brain telling the body, or some combination of the two? Or is it a little bit of both but also something else? Has education and Western science in general unjustly labeled the brain as the command center when in fact it's only one of several interlocking systems that regulate behavior?

At the core of this issue is the science of embodied cognition. This emerging branch of research suggests that cognitive behavior is rooted in the body rather than the brain. That's a simple enough thing to write, but a radical departure from how Western science has explained cognition since at least the Enlightenment. The ramifications are far reaching; they shed light on everything from learning and school to parenting and aging.

Case in point: when babies are born, most possess a series of reflexes that disappear after a few months. These factory-setting, instinctual reactions may serve an evolutionary purpose that humans have moved past in terms of our day-to-day experience of the world. The first is the palmar grasp. This phenomena is observable in babies when someone offers the infant a finger. The baby grasps it, everyone *coos* and *ahs*, and comments are made about what a strong little creature a baby is. Well, yes. It pays to be strong, and quick, particularly when you are an arboreal primate.

Following the river of evolution back a few million years, it is possible to identify tantalizing evidence that early hominids slept in trees. Paleoanthropologists have known for quite some time that at some point around twenty million years ago, the line of hominids that eventually became humans diverged from the common ancestor shared with chimps, bonobos, gorillas, orangutans, and gibbons. It's likely that at the time of the branching of the evolutionary tree, whatever scrubby, hairy ancestor that gave rise to the hominid line that would bequeath the earth to Cardi B and Nicolas Sarkozy slept in trees.

Although twenty million years ago is a fair chunk of time, recent evidence suggests that as recently as three million years ago, hominid ancestors were sleeping in trees. The fossil nicknamed "Selam" was discovered in Ethiopia in 2006. The fossil appears to belong to a young girl of the *Australopithecus afarensis* line of hominids, and Jeremy DeSilva, an anthropology professor at Dartmouth College, believe that the completeness of the skeleton, particularly the foot bones, suggest that human's recent ancestors spent a fair amount of time in trees. DeSilva and colleagues have argued that the modified big toe—not quite the human big toe, not quite the thumb-like big toe of the chimp but rather somewhere in between—suggests a life split between the branches of trees and the grassy plain below.

Many new parents beam with pride at how strong the grasp of their new infant is during the first six months of life, embodying all the desire for a healthy child in that little fist grasping my finger. The truth is that most infants are able to hold their own weight for around ten seconds or even longer just a few weeks after birth. Although there's some debate around the purpose of the palmar grasp, it's entirely possible that it's a vestigial trait born of evolutionary history.

If one is sleeping in a tree and starts to fall out, the ability to instinctively grab Mama's hair, or a branch, and hang on, is a pretty nifty trick and would certainly be conserved as an adaptation. After all, if infants or juveniles can avoid the plunge down to the waiting hyenas or rocky ground below, they've got a much better chance of growing up, wooing a mate, and producing offspring.

Interestingly, there is a similar reflex in the foot of infants known as the *plantar reflex*; stroke a baby's instep or the sole of the foot and the toes curl up. With our stumpy big toes, it's not a very useful adaptation anymore, but with a more grasp-friendly toe like Selam's, that reflex may have helped an infant hang on for dear life as she rode piggyback while her mother dashed away from hungry predators.

The palmar and plantar reflexes are examples of the way in which evolutionary history is encoded not just in how humans look, but how individuals respond to situations, behave, learn, and think. At the core of embodied cognition is this powerful idea that the body's reaction to stimuli is intricately tied to behavioral and cognitive processing and reacting to that stimuli. In some cases, however, science has given the brain A-listing, whereas it may in fact be the body that is doing the heavy lifting in directing behavioral responses. In a surprising number of situations, humans are just like those little babies, hanging on for dear life with no idea why.

SONG OF THE DECAYING BUFFALO CARCASS

Imagine bipedal hominids wandering around east Africa around three to five million years ago, during the formative stages of homo evolution. It's hot and the search for food is on. The hominids are part of a little family band, with maybe ten or twelve individuals total. They're rambling across the savannah, searching for some grub. Suddenly, the hirsute leader whoops gleefully. Troop members follow the sound of the whoop, read the emotion on the leader's face, and then follow her gaze; they see circling vultures. This usually means a carcass—lots of meat and bone and marrow and all manner of nutrient-dense deliciousness to scavenge.

The troop heads over toward the scene. There are a few mangy prehistoric hyenas skulking around, but nothing human's hominid ancestors can't handle with a bit of stick waving and rock throwing. They get up to the carcass—looks like buffalo, but at this stage who can be sure—but then draw up short. The smell hits the troop: putrefaction. It's rotten and no longer fit to eat. The leader's face is downcast. The younger members of the troop, not yet used to the lean times, moan and whimper, and the adults feel that tug of empathy—it's gotta suck to be all excited and then let down when you're a kid. Everyone's hungry, it's hot, and now the troop doesn't have a meal. At this point, the insula of our hominid ancestor is going into overdrive.

The insula is a brain region—two, actually, one in each hemisphere—related to so many different behaviors and functions it's sort of like the Leatherman multitool of the brain. Soon to come is a dive into why the insula is so important, and all the ways it's connected to both how humans feel on the inside and how they perceive the world on the outside, but before that, it's important to note that the insula—or insular cortex if we're being fancy—is still a pretty mysterious part of the brain. Its full functionality, and the way it's connected both to other brain regions and systems, isn't completely understood. What is clear, however, based on countless functional magnetic resonance imaging (fMRI) studies, is that the insula is something of a major crossroads within the interconnected systems of the body, monitoring and assessing both inward states and the environment, as well as directing behavioral responses.

The insula does so much for human beings, it can be a bit dizzying. When an individual looks at someone he or she loves and feels that woozy gush, thank the insula. Crave a slice of chocolate cake with a glass of milk? Insula. In the examination room, waiting for results from the doctor, do you notice the rapid pitter-patter of the heart as it hammers away because it's a bit stressed out? The insula is the part of your brain assessing heart rate.

In fact, the insula is involved in so many different brain functions it can be a bit overwhelming. But one thing is pretty clear: the majority of the functions of this part of the brain can be tied to some sort of evolved adaptation, which is all well and good, but the amazing thing is that it's also tied to some of the more complex human emotional states—and thus behaviors—as well.

As a short list to get started, the insula has been tied to the following functions:

- Perception and processing of sensory input
- Emotional states and feelings
- Motor control
- Decision making
- Risk assessment
- Bodily awareness
- Complex states of self-awareness

Although about the size of a ripe apricot, each of the insula on both sides of the brain seem tied to a large number of functions. Located just below the temporal lobe of the cerebral cortex, among the insula's other noteworthy characteristics is that it contains an unusual type of neuron known as *von Economo cells*, named after the researcher who conclusively identified them in the 1920s, Constantin von Economo. The function of these cells is not entirely understood, but it seems likely that the little buggers have something to do with complex emotional states and social skills.

A helpful way to think about the insula is as the connective hub of the brain and body—the switchboard between our senses and perception and our interoceptive awareness. In short, it mediates how we feel on the inside and out.

Information from the environment—whether it's a warm, southerly breeze; the wafting scent of fresh baked bread; a charging hippopotamus; or the barking, disgruntled voice of an employee's manager—gets funneled through the insula. In addition, the insula is wired to the interoceptive network, monitoring body temperature, heart rate, the motility of the digestive system, and the body's homeostasis or lack thereof.

Although there are specific regions of the insula that respond to specific sensory input, there's enough cross-networking that it's helpful to think of the insula as the mass of wires behind your desktop computer: lots of crisscrossed, tangled cords connecting far-flung systems. In addition, the insula is connected to the prefrontal cortex, long thought to be the home of executive functions and complex thinking—cognition, in a nutshell.

The insula both receives input from the body's various interoceptive sites and directs some involuntary processes such as blood pressure. Body states such as hunger, pain, thirst, itchiness, and others are channeled through the insula. And here's where human's history as foraging primates comes more clearly into focus. When individuals perceive certain sounds, or pictures even, that relate to some of these bodily states, the insula reacts accordingly. That's why a picture of a cheesecake makes people just as hungry as actually seeing a cheesecake. This is the essence of the embodied nature of experience: it is not just *thinking* one is hungry for cheesecake when looking at the picture. As far as the experience of the body is concerned, one *is* hungry for cheesecake. There's a deliciously crunchy graham cracker crust and a thick, luxurious cheesecake filling. The fork slices down, and the happy diner hefts the creamy wedge toward the mouth. . . . Even reading those words can elicit an embodied reaction. If you're a cheesecake fan (and if you're not, feel free to substitute your crave-worthy dish of choice) and you're simply *reading* about cheesecake, your body processes that symbolic information as the real thing. The connectivity between the prefrontal cortex and interoceptive awareness, through the insula, underpins the way in which the experience of cognition itself is embodied.

But because of the insula's connection not just to the prefrontal cortex but to other regions such as the limbic system, particularly the amygdala, it plays a part in other processes as well. The role of this little blob of gray matter known as the insula goes beyond the love of desserts. For example, fear is processed, in part, in the insula—long thought to be the property of the amygdala. Other emotions such as disgust, happiness, sadness, surprise, and sheer joy are processed in part through the hard-working insula. The insula plays a pivotal role in one other aspect of human behavior of particular note: the ability to learn.

THE INSULA AT THE BACK OF THE CLASS

In an article for *Current Neurology and Neuroscience Reports*, Nadine Gogolla of the Max Planck Institute of Neurobiology summed up the role of the insula in learning: "The insular cortex also mediates long-term retention of appetitive, aversive, or novelty-driven learning."[1]

This is not a trivial statement, as the way in which we learn has long been thought—from the cognitive psychologists of the mid-twentieth century all the way back to Socrates—as a purely disembodied process. There have been plenty of competing ideologies from various schools of philosophy, education, and science, such as John Dewey's belief in experiential learning, but for the most part learning more complex, abstract information was seen as a higher order of thought and processing that was different than, say, learning to avoid touching a hot stove or eating gross stuff off the ground (five-second rule notwithstanding).

Here are a few terms to help bushwhack through this next bit. *Salience* describes the quality of "novelty" of a given environmental stimuli. Most animals—rats in labs, but also chimps in zoos, toddlers in experiments, and dogs just about everywhere—respond more acutely and with a greater level of exploratory behavior to new things in their environment. Particularly when they're young, dogs with an affinity for humans like Labs and golden retrievers will practically bowl over new visitors to the home, and, although the owners still get a hearty wag and sniff when they arrive, it is clear that a new person means something special to the dog.

Put a bunch of rats in an enclosure and one night when they're all sleeping, put a red ball in the corner of their cage; chances are they'll all take a good sniff and look at this new weird thing come morning. This phenomena is crystal clear to anyone who celebrates Christmas and observes their kids when they open stockings—new stuff! Novelty! It makes sense on an evolutionary scale. New things—smells, tastes, objects, people—in the environment can mean new sources of food, mates, and resources. Evolving a certain eagerness to check out the unfamiliar is an adaptation that most animals have.

The *valence* of a given environment or object is its value on the good/bad scale. Assessing valence is a crucial evolutionary skill. A leaf that tastes intensely bitter could be poison. Judge a brooding alpha male's temper; maybe he's lonely or he's looking to club someone over the head to cheer himself up. The given valence and salience of objects, situations, and individuals in the environment have everything to do with the insula and the way individuals learn.

It works like this. Imagine an early hominid ancestor cruising around the plains of east Africa looking for a bite to eat. A bush is spotted. It's unfamiliar, with little clusters of berries that this representative sample of human's naked ancestors has never seen. Next to it is a familiar food source, maybe some tough, stringy ol' tubers that, yes, satisfy the rumble in her tummy, but are about as much fun to eat as a bath mat. Human's primitive ancestor approaches the funny new bush with the little red berries. A quick glance at the tuber doesn't spur much in the insula in terms of reaction. This tuber is a familiar, nonnovel item in the environment. She's miserably chewed her way through hundreds of those. But the berries rate high in salience, meaning they are brand-spanking-new to her.

She plucks one and places it in her mouth, crunching the red fruit. Ew! It's bitter, gross, and tastes like a hyena's butt. She spits it out, but the flavor—bitter, nasty, and clearly meant to be avoided—remains. During this moment, her insula fires up and releases acetylcholine, a neurotransmitter instrumental in pain responses. When muscles contract, as they do in moments of pain aversion, acetylcholine is produced. Thus, the insula plays a pretty big role in learning about our environment—what's good, what's bad, and what's familiar. This is crucial learning, especially as the insula responds not only to crappy-tasting berries but to situations, people, and images. That means that when a situation is unpleasant and unfamiliar, it's possible that the insula is in overdrive, encoding that particular memory as one to be avoided, sort of like on the first day of middle school when mom made you wear that god-awful sweater. Blech.

However, it's not all bad news from the insula. Looking at those berries prior to eating them spurred a dopaminergic response in our ancestor's brain. Dopamine—the brain's all-purpose reward system—gets released upon investigating novel and ambiguous environmental stimuli. Humans have adapted to enjoy, in the neurochemical sense, exploring new and unknown things. There's a limit, of course, to how funky and weird and unknown a thing can be before individuals skedaddle out of there, but a little novelty goes a long way.

In a nutshell, learning and memory, which are both central pillars of education, are dependent on the work of this part of the brain known as the insula. And the connections between interoceptive awareness, sensorimotor perception, and the ability to remember new information are all interconnected through these little apricot-sized bits of brain matter. Memory itself depends on it, whether it's remembering the capitals of all fifty states or knowing where the best tree is to escape a hungry lion.

THE HOLY CONUNDRUM OF COOL RANCH DORITOS

Education in general has a tendency to think of decisions like a Sherlock Holmes narrative. Individuals rule out the preposterous and impossible, weigh the evidence, and come up with a probable answer or solution. But this definition of decision making leaves out the body. The Sherlockian process is a purely rational act, and doesn't take into account the embodied nature of experience whatsoever. When a student makes a decision of what class to take, or how to answer a particular essay question, or how they "feel" about a subject, they are depending on a deeply embodied experience that involves—no surprise here—the insula.

A student is considering whether to go to the library and knock out another hour of studying for the dreaded accounting exam, or head over and see who's in the common room and whether or not there may be some mind-altering substances like Fortnite or Cool Ranch Doritos on hand to while away the rest of a rainy Saturday. This decision is usually considered in cognitive terms, weighing the perceived benefits of being more prepared for the test against enjoying some free time with friends. And while this cognitive process of rationality (or lack thereof) is a psychological one, it's also very much embodied.

Remember *valence*, the good-versus-bad scale that depends on the insula? Well, in this example, the insula is hard at work balancing the valence, importance, and likelihood of multiple outcomes during this moment of scholarly waffling. But remember: *we all have a body*. The body is a major player in this moment. Antonio Damasio, the brilliant forefather of much of embodied cognition research, believed that the body and its needs, wants, and motivations were deeply embedded in all decisions. He called this idea his "somatic marker hypothesis."[2] In this example, it's not just the rational, cognitive situation at work, but also emotions and the body ("gut feelings") that are operating on the student's ability to make a choice.

If decisions are made as much by the body as by the brain, then what part of the brain is a vital connector between the ways in which those two act in concert? The part that responds to both sensorimotor stimuli (the environment), interoceptive information (how a person "feels"), and novelty and memory, salience and valence.

To make it plain: decisions live in the insula.

Commonly, when the act of making decisions is considered, it is thought that the brain is making a good choice when it's often the body that has taken the reins. Part of the role dopamine has in the insula is in prediction. Anticipating bodily rewards—don't forget about those delicious, crunchy Cool Ranch Doritos—can change the way neurons fire within the insula during the decision-making process. That means the insula may be guiding the student toward the potentiality of those Doritos, even though they think they're just taking a study break.

So, the student tosses his backpack on his shoulder and heads off to the common room, already subconsciously dreaming of mainlining Skittles and zoning out playing video games. As he rambles across the frozen campus, he sees a fellow student slip on some ice with an armload of books and come down, hard, on her tailbone. It's that sort of fall that reverberates through the skeleton, a dull, fleshy thump that can be heard from twenty feet away.

The insula, which was been busy interfering with our student's desire to be an accountant in the next four years, now also offers something a bit more heartening: empathy. Just as the insula was vital in the processing of the negative valence of the nasty berries our Pleistocene ancestor ate earlier, it now activates while simply *watching* someone experience pain. This suggests that among the many aspects of behavior that are related to the insula, empathy may be one of the most important. After all, as social primates, if someone else's pain causes pain, than the easing of that pain and suffering is an excellent strategy to bank some care and concern. That's a handy evolutionary adaptation if one depends on the herd for protection and survival. It also means that abstract definitions of suffering, or theory-heavy descriptions of inequity, may not do the trick. But direct exposure to people in pain can activate the neural home for empathy, and may even spur change.

THE APRICOT THAT DRIVES THE BUS

The importance of the insula in behavior can't be overstated. A huge amount of neurobiological research has focused on the prefrontal cortex over the last fifty years. This makes sense. It's the home of our executive functions, and it's smart and sexy and witty. It's our brain's Ali Wong or Trevor Noah. In addition, it's easy to see and manipulate physically—it's the top layer just under the skull. But with advances in fMRI, it's becoming possible to read the deeper—and older—layers of the brain and see what they're up to. That's the insula.

So what does this all mean? To recap:

- The insula connects the body's perception of the world around us through the senses to our brain.
- Interoceptive awareness is connected directly to the insula.
- The role the insula plays in memory and learning is based on structures that are embodied through the senses. Think of that unpleasant berry eaten by the hominid ancestor as the same thing as that teacher everyone avoids taking because he's wicked creepy and mean.
- When individuals make decisions, although they may not check in with their bodies, the insula sure gets a vote and it becomes part of the decision-making process.
- Exploring new stuff is awesome. Familiar stuff is kind of boring.
- Learning, memory, decision making, and empathy are all grounded in an evolutionary older part of the brain that also serves as the neural substrate for the sense of embodiment and the perception of the world through the senses.

Education in general is based on an understanding of learning that comes from the Enlightenment-era ideas of Descartes. Embodiment rarely comes into conversations. Researchers even tend to discuss learning disabilities like attention deficit hyperactivity disorder (ADHD) in purely psychological or neurochemical terms, despite the growing understanding of the role the body plays in behavior and decision making.

Humans' sensorimotor perception of the world—its sights, smells, tastes, tactile sensations, and sounds—are the foundation for the body's and brain's structuring. The first way humans learned about the world, both as infants and, on a species level, as hominids, was through the incorporation of environmental stimuli and interoceptive feedback. And it is still so, despite institutions that deny children and young adults their embodied natures.

CHOPSTICK MASTERY

To understand learning, it's helpful to explore the neuroscience behind how individuals remember information, process experience, and change behavior. This is no easy task; the connection between learning, behavior, and the embodied biology of the way in which organisms interact with their environment is incredibly complex. A slow, steady approach is best.

The main point of all of this is simply that acquiring new skills, information, or behaviors (i.e., learning) is not some abstract, purely rational consequence of a disembodied mind going about its business on some ethereal plane. Rather, learning is very much an embodied, physical process. It can be measured (to a certain extent) in neurons, gray matter, and synaptic connectivity. Learning happens in, and to, the body. This is something that early twentieth-century educational philosopher John Dewey knew well, only now there are fMRI machines to explain why it's happening.

Human bodies learn both physically and abstractly. The two are intimately tied and intertwined, however, and talking about them separately works only up to a point. Once the underlying endocrinology, behavioral dynamic, and neurobiology is explored, the two types of learning—shooting three-pointers and lecturing on quantum mechanics—are actually remarkably similar.

For the sake of clarity, however, let's look at purely physical learning. This is an easy concept for any adult who has tried to use chopsticks for the first time. At first, the two skinny sticks seem incompatible with the consumption of food. They are unwieldy and awkward; balancing the two chopsticks in such a way as to have a pincer-like tool capable of hoisting up a dumpling or California roll seems impossible. How the heck are you supposed to do this with only one hand?

The adult novice chopstick user uses different combinations, rearranging the chopsticks in different ways, training the muscles of the hand and fingers until, slowly, after many hours spent engaged in a sort of masochistic, self-imposed Pavlovian drool-experiment, it becomes possible to sit at dinner and not risk starving to death, or at least avoid spilling rice down your shirtfront to pile up, uneaten, in your lap. But the physical acquisition of this type of skill takes time and repetition and experimentation.

Ever tried describing to someone, without visual cues, how to hold chopsticks? It can be done, but it's way easier to just grab some and figure it out. This learning, popularly called "muscle memory" (a wonderful phrase that suggests that an understanding of embodied cognition has infiltrated our common expressions) infuses our hands and fingers until they just seem to eventually "get it" and go about the business of carting gyoza and pad thai into our gullets.

Abstract concepts are rather the same in terms of how the brain interprets them and stores them as memory. The acquisition and mastery of an idea or concept (terms that can be used interchangeably) follows a similar trajectory. People learn about the concept of, for example, democracy. Like the initial forays in feeding with tiny, slender bamboo shoots at the local dim sum joint, an understanding of democracy is crude and simple at first, and not particularly useful. It's only after repeated attempts to understand the way democracy works, and participating in democratic experiences, that individuals can begin to understand the role voting, representation, checks and balances, majorities, and argumentation play in a functional democracy. Mastery of the concept comes when someone can fluidly both use and critique democracy (just as experienced chopstick users can eat and carry on a conversation and flag down the waiter for another beer without losing any food), see its limitations and loopholes, and see that what is popular in a democracy is not always beneficial to the electorate.

The thing to remember is that both these types of learning—mastering ideas and mastering skills—use the same neural pathways. Not getting lost in the woods by recognizing landmarks isn't terribly different from not getting lost in the plot of *Anna Karenina*. The reason is that the structures, which initially evolved in our bodies and brains to negotiate the physical landscape on a spatial basis, are recruited to navigate text.

Many educational goals are based on this ability that lies in the bedrock of embodied experience. Developing a narrative-based argument, for example, follows the same structural processes as following a complex plot. Keeping track of a quadratic equation is remarkably similar to following a line of thought; if important points are left out, the ending doesn't make sense. This basic understanding of cause and effect and the consequences of actions are hardwired into primate behavior. Even something as simple as hearing thunder and hightailing it for some shelter is based in this anticipatory cognitive process. When educators offer young students the chance to first develop the neural systems and embodied functions of memory used in negotiating unstructured landscapes, those students will develop the capacity to use a more robust embodied cognitive approach to traditional problems and lessons in education.

For example, the way in which landmarks get remembered and the environment navigated is connected to a process called *spatial reasoning*. It's written that Einstein had an ability to see objects in his mind and rotate them, move them through mental space this way and that to figure out how the universe worked.[3] His highly developed spatial reasoning skills are related to the way in which some basketball players such as Magic Johnson have "court sense," the ability to perceive everything that is happening around them during a game rather than just having tunnel vision.

This can be seen in the classroom, particularly with students who struggle with attention disorders. In fact, it's been shown that individuals with ADHD lack the ability to perceive their environment holistically compared with non-ADHD folks. *Global precedence* is the term used to apply to the ability to see the large-scale picture rather than more localized information. Being able to discern the large-scale patterns versus getting hung up on irrelevant details is crucial, particularly in current education structures. This isn't to say that details aren't important—they are—but when a student is struggling with finding the right font for their essay, or can't figure out how to write a three-page essay because the overall narrative structure and argument is unclear, it likely has something to do with the level of their global precedence.

In fact, exercise has been shown to help treat ADHD. The reasons behind this are complex, but it seems likely that physical exercise—activities that usually require moving through space and being aware of surroundings in a macro versus micro sense—is connected to the development of more acute global precedence. So from a practical perspective, if a student is struggling with writing or math, it's not inconceivable that creating large chunks of time engaged in physical, outdoor playtime in complex environments (the woods, the jungle gym, the vacant lot, the unused gully behind the school) may play a role in mitigating the negative effects of atrophied global precedence. In short, might as well let the kids go play in the trees like monkeys because that's basically what they are.

LEARNING LUCY

Okay, not *actually* monkeys. Monkeys and humans had a common ancestor about twenty-five million years ago. Since that time, the respective species have changed considerably, but there's no doubt that as two representative primates, humans and monkeys as well as apes are pretty similar, particularly on a neurological and behavioral level. This is as true for macaques as it is for tamarins, Chicagoans, and Cambodians. A primate is a primate, broadly speaking—not identical, but similar.

This is the first stop at the evolution kiosk of *why learning is embodied*. Learning is embodied because of the environmental context in which primates (humans) evolved over millennia on a biological as well as behavioral level. The environment hominids negotiated over the past three million years, and the way bodies evolved to exist in that environment, are directly related to the behaviors developed to ensure success as a species. And whatever happened on the African savannah during that time must've been okay, since humans are still here and not extinct, despite the evidence that people are working like crazy to speed up the whole extinction thing by burning up our own planet.

In his 2017 book *Behave*, Robert Sapolsky of Stanford elegantly outlines an argument for why humans act the way they do, and provides a structure for how to backtrack behaviors to get a clearer picture of why humans do some of the crazy, and not-so-crazy, things they do.[4] The equation, according to Sapolsky, goes something like this:

Evolution > Environment > Endocrinology > Neurobiology = Behavior

Before the neuroscientists out there have an aneurysm, it's vital to note here that this is a simplified version of Sapolsky's explanation—I'd suggest that science enthusiasts should grab a copy of his book. It's brilliant. But this summary will do for our purposes of looking at the embodied nature of learning.

Learning is embodied for a number of reasons, but mostly because humans *have* bodies, which seems like a "duh, obviously" sort of explanation, but it's important precisely because of its implicit foundational significance. It is easy to forget, all the time, that everything individuals do or think or see or experience happens through the body, the omnipresent frame of reference. It was through the body's interaction with the environment over the course of the past three million years that humans developed the brains that are now used to vote on which American has talent, consider such concepts as dark matter and socialized medicine, and calculate the volume of a sphere. By the way, "three million years ago" is being used as a general time frame simply because it was about three million years ago that one of our earliest hominid ancestors—the australopithecine species made famous by the fossil named "Lucy" discovered by paleoanthropologist Donald Johanson in 1974—lived in Africa.

One could use five million years just as easily, as other hominid remains like "Ardi" from Ethiopia keep pushing the time line back, but for the sake of keeping things simple, this argument uses the time frame of three million years. However, many of the embodied structures discussed occur in mice and turkeys and tuna as well, and are even older than five million years, evolutionarily speaking. Homo sapiens' relevant evolutionary changes start cropping up during that three-million-year span, especially the part of the brain known as the *prefrontal cortex*. But that's a whole other tangent. Back to Lucy.

So there were humanity's ancestors, little 100-pound, ape-like creatures, bopping around Africa about three million years ago. To ensure that they didn't get eaten or starve or freeze or drown or get trampled, they had to learn. Although their brains were already pretty big, it was really the body that did most of the learning, namely through the way in which senses collected information about the environment and the manner in which the amygdala processed that information and helped steer behaviors.

A quick note here on the complicated nature of the interactions described in this chapter. The neural structures and activities that link evolution and behavior are incredibly complex. Reading through the articles practically requires a researcher to be both a scientific linguist and an interpreter of scholar-talk. In the interest of keeping the reader awake, this book has tried to make things a bit more digestible and simple. But that means that what is being described is done so in general terms.

For every rule neuroscientists seem to come up with about the relationship between cognition and the brain, or the body for that matter, some other group of neuroscientists discover an exception or a different angle. As a corollary, the second I say, "Tall people are more likely to dunk basketballs," you'll have legions of Spud Webb (5'7") and Nate Robinson (5'9") fans howling that I'm a liar. Okay, you've been warned. Back to the savannah.

So, out there in Africa about three million years ago, humans' hominid ancestors had to learn some pretty important stuff to stay alive, locate food, have offspring, and contribute to their Paleolithic 401(k)s. They were able to do this in large part thanks to a section of the brain known as the *limbic system*. The word *limbic* just means "border," and it's kind of a boring name for this amazing part of the brain. Recently, some scientists have referred to the limbic system as the "paleomammalian cortex," which sounds kind of edgy and cool. Perhaps for brevity's sake, *limbic* will work. It is also easier to spell.

When humans' hominid ancestors were locating water and food and avoiding saber-toothed tigers, they did so by depending on visual, olfactory, auditory, haptic, and interoceptive inputs. When humans receive information from the senses, these types of stimuli are plugged right into the amygdala—part of the limbic system. The amygdala is a blob of neurons that sits underneath the big dome of the cortex and, with other structures such as the hypothalamus and hippocampus, help process environmental stimuli received from the senses, internal information like hunger and thirst, and then turn these experiences from short-term memories into long-term memories.

From an evolutionary perspective, this is what learning looked like early in hominid development. It's important to keep in mind, however, that this is the same sort of learning that plenty of other animals exhibit. The human limbic system—via the amygdala and hippocampus—"learns" to associate the smell of, for example, an angry alpha male with fear, since every time the alpha gets pissed off, he starts walloping everyone around him.

Another important note here about the way neural inputs into the amygdala work: they work fast. This is the emotional, knee-jerk, instinctive part of human behavior. It's the amygdala that has individuals leaping back from the pile of rocks they happen to be climbing over on a hike because they've heard a "hiss," only to discover that the sound was merely a cascading stream of pebbles and not a deadly serpent. When behavior is informed by the amygdala, the benefit is that the response happens in a blink. The drawback is that the amygdala makes mistakes. A bouncing shadow at night becomes a bat seeking to entangle itself in someone's hair and everyone screams and flaps their hands about ridiculously, when it's actually just a moth fluttering around by the porch light.

Okay, time to circle the wagons. The point here is this: embodied cognition as a concept that informs learning rests on the understanding that the very way humans are embodied is through the senses. Thus how individuals process sensorimotor information in the brain is important. The place this happens is the limbic system, primarily the amygdala and the hypothalamus, which acts as the body's thermostat and tries to keep the body in homeostasis (not too cold, hot, hungry, or thirsty).

To put all this into simpler, more rodent-friendly terms: rats dedicate a huge proportion of their neurons to the olfactory inputs they receive. They navigate the world primarily through smell, so this makes perfect sense. Rats dedicate about 40 percent of certain brain fibers to the olfactory department. How bodies interact with the environment, via the senses, bears directly on the brain's cognitive abilities. It's true for Mickey, Minnie, Mighty, and humans.

What this means on a practical level is surprisingly simple. Humans evolved to figure out problems in an embodied way, navigating through the world. Navigating modern problems—tax returns, social studies quizzes, SATs, and relationships—relies on the same physical structures. So a school that prioritizes outdoor, physical activity, especially in unstructured environments, from preschool through college, is directly nurturing the embodied systems that help make good choices and quick decisions and that solve problems. Denying those types of experiences by increasing class time, keeping children indoors, and limiting access to physical activities may inhibit children's ability to learn.

Recess anyone?

NOTES

1. Nadine Gogolla, "The Insular Cortex." *Current Neurology and Neuroscience Reports* (June 19, 2017), www.ncbi.nlm.nih.gov/pubmed/28633023.

2. Antonio Damasio, "The Somatic Marker Hypothesis and the Possible Functions of the Prefrontal Cortex." *The Prefrontal CortexExecutive and Cognitive Functions* (Apr. 1998), 36–50, doi:10.1093/acprof:oso/9780198524410.003.0004.

3. Walter Isaacson, *Einstein: His Life and Universe* (New York: Simon & Schuster, 2008).

4. Robert Sapolsky, *Behave: The Biology of Humans at Our Best and Worst* (New York: Vintage, 2018).

FOUR

Got Dopamine?

What does learning look like? Supposing someone took everything they learned over the course of the past year and dumped it out on the table. What would they have? It wouldn't necessarily be symbolic representations of learning, such as a copy of Ralph Ellison's *Invisible Man* on the table because they read that particular novel, or a list of state capitals on a piece of paper. In this case, what represents the tangible, physical, embodied dividends of learning? Because if learning is truly embodied, it must be in the body, made of stuff—tissue, bone, muscle, and blood. If learning cannot be represented physically, as an embodied reality, then maybe Descartes was right after all?

Only he wasn't. Here's a similar question: is it possible to record, physically, all the sorrow one has felt in the past year? Well, sure, if one could somehow save all the tears one has cried in a bucket, they could be weighed. It would be a quantifiable way to figure out something as amorphous as sorrow. "I've had sixteen fluid ounces of shitty things happen this past year," you could brag.

Perhaps it's time to double down and provide evidence that learning is embodied by talking about stuff. To do so, it makes sense to focus on two parts of neurobiology that play a major role in learning: dopamine and synaptic pathways. This is going to get a bit science-y, but once you've read through this part you'll be able to impress your friends at dinner parties with lines like, "Wow, Janet, those canapés really fired up my mesolimbic dopaminergic reward system!" You'll be super popular. Honest.

YOUR FRIENDLY DOPAMINE PRIMER

Dopamine is a neurotransmitter in the body directly related to rewards. It's what the body produces when it feels good or euphoric. It's synthesized in various locations, one of which is the mesolimbic and mesocortical dopamine system, namely, dopamine that affects our limbic system and cortex.

Dopamine's role in human behavior and the relationship it has to various other interconnected systems underlies a significant part of behavior. There are a few things pertinent to dopamine production and learning that will be helpful to understand.

1. **Dopamine is the body's evolved response to pleasurable stimuli.** Archaic humans produced dopamine when they were scavenging around the savannah and spotted some delicious berries, juicy and ripe on the bush. Dopamine surged as they anticipated the sweet, plump goodness of those berries. Over time, this is a good thing in evolutionary terms: connecting nutrient-rich foods with a pleasurable body state is an advantageous strategy for survival.
2. **Dopamine feels good.** The release of dopamine in the body is pleasurable. It fills human beings with tingly goodness. Some activities that release dopamine are sex, exercise, positive social interactions, the sight of good food, and even acts of generosity. However, dopamine is also produced through drug use, calorically dense and sugary foods, and even punishing someone.
3. **The more dopamine you get, the more you need.** Humans are never exactly sated when it comes to dopamine—the body always craves more, more, more. In addition, as time goes on, the body can demand bigger and bigger doses. This is, in part, why drug users overdose. They take bigger and bigger hits of heroin because they want more and more dopamine released.
4. **Our dopamine system can be blunted.** Just like heroin addicts who wreak havoc on their dopamine reward system by overindulging, other stimuli can blunt or reduce the excitability of dopamine production.
5. **Dopamine is produced during anticipatory states.** This last little nugget of dinner-table witticism is something that is radically interesting and is potentially tied to how we learn.

LEARNING IS A GAMBLE

Ideally, school would leverage the dopaminergic reward system to help students excel. For example, get an "A" on a test and suddenly *squirt!* Dopamine is released in the brain, the student feels euphoric, and connects this good feeling with high test scores and *voila!*—they're summa cum laude. And while it's likely that students who get good grades could be driven by a quest for dopamine, the way dopamine interacts with bodies and behavior is a bit more nuanced in incredibly important ways.

It works like this: I take my dog outside for some exercise, and decide to throw a tennis ball for her to fetch. Because she's a capricious and goofy-as-hell golden retriever, sometimes when she runs off to grab the ball, she decides to switch games and play keep away, running about, tail wagging, unwilling to give up the ball without a joyful tug-of-war. But, since I'm someone with a prefrontal cortex honed by years of watching Cesar Millan on the Discovery Channel, I offer the dog a treat when she returns the ball and drops it at my feet. Doggy dopamine effect achieved.

Here's where things get interesting. Dopamine production starts during the task in question—maybe even when we head outside and the dog sees a tennis ball in my hand. *"Oh boy oh boy oh boy! Erik is going to play with me and give me treats!"* thinks my dog. Then I chuck the ball. It is sometime around this period—the anticipatory period—that dopamine production is at its highest. Mind you, I have yet to provide the reward. The dog has learned, at this point in our daily romps, a reward is coming. By the time she actually chomps the biscuit I've given her, dopamine production has dropped well below what it was before. The reward is never as satisfying—on a dopaminergic level—as the *anticipation* of that reward.

But it gets even more intense if I decide to only reward her once in a while for dropping the ball at my feet. This uncertainty—ambiguity, if you will—ramps up dopamine production *even more* in my dog. It's the not knowing if she'll be rewarded balanced with the memory of reward in similar circumstances that causes the behavior.

Dopamine production is directly related to learning. Individuals get the biggest dopamine burst when they know *something* great is coming, but aren't sure *what*, or *when*. Humans are all, at a certain level, wired for this game of odds—gamblers in life if not in the casino.

To align educational practices with this understanding, schools would have to acknowledge that when students gain mastery of a subject, or understanding of a concept, they are actually embodying this knowledge through the production of dopamine. Learning has a physical effect, or learning is physical. Student behaviors in school aren't only driven by arbitrary goals like grades or prestige or the opportunity to get on the dean's list or the desire for a Maserati. It's the body that gives them the impetus to work toward figuring out that pesky equation, the body that provides motivation to plow through the dense assigned reading, and the body that motivates students in the lab to try the experiment. One. More. Time.

One last thing about dopamine. As schools race to provide more exciting educational technology (edtech) experiences for students, all the bright lights, bells, and whistles of the digital medium only serve to habituate students to stimuli. It deadens their dopaminergic responses, decreasing the way the body primes them for learning. This isn't to say that edtech is like heroin or cocaine in its ability to dampen dopamine responses. In fact, much of edtech—which is beginning more and more to resemble social media interfaces—is designed explicitly to engage the dopaminergic system. Because dopamine reinforces behaviors, it's no wonder that many digital platforms play a similar game with users as I played with my dog. Case in point: Instagram will withhold "likes" on a post, then dump all the likes in a big chunk. This keeps us checking our phones, hoping for that randomized reward. Edtech programs include learning management systems such as Canvas, the online system developed by Instructure. At the most basic level, Canvas adopts a model similar to Instagram, where students can constantly check their phone app to see when and if an assignment has been graded. No need here for an algorithm that randomizes what a student sees and when; capricious instructors take care of that part.

In the pragmatic, hectic world of education, it makes sense to have a software program to organize assignments and readings and assessment and attendance and grades. However, these systems share many features (and in some cases are financially linked) to social media and consumer-oriented sites. Perhaps teachers may want to consider the way in which their students are engaging with their learning through technologies designed to get them hooked on their screens. A simple remedy is to offer real feedback, face to face. Another is to schedule grading in such a way that it's not randomly timed. All this requires work, of course. But if the goal of edtech is to maximize screen touches and interactions, then it is working directly against the goals of teachers, which is to create an environment in which deliberation, contemplation, and personal development take the fore.

TOTALLY RIPPED BRAINS

Okay, now on to synaptic connections! Learning makes our brain stronger. But how does that happen, and what, exactly, does "stronger" mean? It's commonly understood how muscles get big: someone who is working out is putting repeated stress and load on the muscle cells, a stimulus that causes the muscle cells to divide more than usual to create more muscle until the person can run down the beach dressed in only a swimsuit, causing hearts to flutter. Does the brain strengthen in a similar way? The answer is neither yes nor no, because the brain—although made up of cells like everything else in the body—doesn't create new cells. Or does it? It does, just not a lot and in strange ways.

All right, now that's confusing. To straighten things out, it's helpful to define what the brain is actually made of.

The brain is made up of cells (neurons). There are also cells in the brain called "glia cells," which function as support for the system, but it's the neurons that do the stuff that pertains most directly to learning.

To imagine what neurons look like, picture a basic stick-figure drawing of a human. The head is the cell body of the neuron with the nucleus that students learned about in tenth-grade biology. Now, draw some crazy hair sticking out all over the figure's head. Those are dendrites, filaments that receive information from other cells throughout the nervous system. The elongated body of the stick figure neuron is the axon, and instead of legs and feet we draw what looks like tree roots—those are the nerve endings. Basically, a neuron looks like a Rastafarian Ent from a J. R. R. Tolkien fantasy world.

When the brain learns information, it "strengthens" the synaptic connection between neurons. This process is incredibly complex, involving the neurotransmitter glutamate and special receptors in the dendrites. But in essence, as humans repeatedly experience the same thing, the synaptic connection becomes more tightly coupled, and with enough repetition individuals have a memory of that information. This is how experiences change the structure of our brains. Each human being on the planet has all these synaptic connections that have been designed through the life he or she has led. A master aikido practitioner's synaptic connections are different from an opera singer's, whose neurons are coupled in different ways than a spelling bee champion's.

Now, without getting overly complex and sending the reader off on the next train to Napville, those synaptic connections turn into memory when that information is held onto through a process called *long-term potentiation* and *long-term depression*. *Long-term potentiation* describes how tighter synaptic coupling happens in the hippocampus—the place the brain stores memories. The second, *long-term depression*, also occurs in the hippocampus, but again, this gets so technical that it'll melt dendrites, so in simpler terms, the long-term depression serves to "prune" the memory in such a way that superfluous information gets excised. Humans dump what is not needed in memories. For example, an individual doesn't remember every single day of third grade, but can return to their school and find the classroom just fine.

One last amazing thing about how learning happens in relation to memory is that it's not *only* in the hippocampus where there is this type of neurogenesis, or new cell manufacture, occurs. It happens *throughout the entire body*. That's right; this whole long-term potentiation thing, which is the basic cellular composition of learning and memory, is *embodied*.

We're in the final stretch now. The brain doesn't just get uniquely wired because of individual experiences. Human brains can also grow *new* neurons. Yep—humans can grow bigger and more robust brains. This happens, as mentioned, primarily in the hippocampus, the part of the limbic system dedicated to memories, and in particular helps differentiate between similar objects within the same class. For example, what's the difference between an omelet and a frittata? Both are egg dishes, but with characteristic differences, as the omelet folds over the filling, and the frittata has it all mixed in together. See? If a person reading that information didn't know the difference before, they do now, and their hippocampus just got bigger.

Back to the initial question: if learning is embodied, a physical, tangible process, what does it look like? Well, if the learning an individual experienced was dumped out on a table, there'd be some dopamine, some nicely coupled synapses, and maybe some neurons from the hippocampus. Obviously, the piles would be really small volumes here—atoms and molecule-type stuff—but still. There it is.

Learning is not some detached, disembodied thing floating around in the ether. It's in the body.

FIVE
The DeMoN in Our Heads

> In daily life, we are often lost in thought. We get lost in regrets about the past and fears about the future. We get lost in our plans, our anger, and our anxiety. At such moments, we cannot really be here for ourselves. We are not really here for life.
> —Thich Nhat Hanh, from *You Are Here*[1]

It starts early, maybe in the shower as a student kicks her day into gear. Perhaps even before that, as she rises out of sleep, it begins. The voice in her head. Nagging, nonsensical. Whinging. Maybe even hysterical, or perhaps whispering in her ear that she's just not good enough. The voice insists that *you're not worthy,* only it's not funny like when Wayne and Garth say it. It's chaos, really. Thoughts circle about wildly, looping around in anxious circles, interrupted by random snippets of coherent thought tangled with song lyrics. "Maroon 5, what the hell are you doing in my head?" our young learner desperately asks.

As she shovels cereal in, or maybe when she's finally on the curb, waiting for the bus, or perhaps when she's battling her way through the streets to make it to school on time, the voices get more insistent. Things she forgot to do. The way she yelled at her mother the night before. The argument she had on social media with a friend replaying in the cinema of her head, endlessly looping through what was said and imagining different scenarios, editing memory and circumstance to say the right things—reimagining victory or a successful negotiation for a cease-fire. Thoughts go caroming around her skull in an uncontrollable stampede.

Her heart quickens. Sweat dampens her pits. Knuckles whiten. Teeth grind.

The DeMoN has taken control.

The **default mode network**—we'll call it the DeMoN—is the collection of neural pathways and connections that activate when the brain is at a state of rest and not cognitively engaged in some task. For the scientifically minded, the DeMoN is a network that connects the medial prefrontal cortex, the posterior cingulate/precuneus, and the angular gyrus (plus other cortices such as the temporal lobe). These parts of the brain play roles in many functions, notably balancing attention between the environment (the external world) and interiority of the body (our internal state of mind), as well as impulse control and emotional regulation. All of the actions in these parts of the brain are regulated by neurotransmitters, in this case dopamine and serotonin.

Put another way, these parts of the brain have been shown to ping and activate when the body is not concentrated on some deliberate cognitive or sensorimotor task. And, usually, for many individuals, the state of being's engagement with the default mode network can be stressful; it's a little kooky. There can be self-loathing, anxiety, and other negative thought patterns. Like wild horses, thoughts gallop heedlessly this way and that. It can, at times, take a Herculean effort to round them up.

In addition, there is a physical corollary. When the focus of the brain is nowhere to be found and the default mode network is running the show, stress often manifests physically. Thinking is not productive or solution oriented. Thought patterns are disjointed and chaotic. This can even be true for individuals who appear to be calm and relaxed; often that is an exterior posture designed and called into play to cover the anxiety-producing effects of the DeMoN.

PROFILING THE DEMON

So what, exactly, does this mean, and how does it play out in the daily life of the individual who is adversely affected by the default mode network? What is this DeMoN in the head, and how did it get there? The default mode network is made of the parts of the brain that activate when thoughts are directed inward, rather than outward toward the environment. For example, if a first-year high school student is carefully picking her way up a rocky ridge to climb a mountain on a class trip, her DeMoN is probably pretty quiet—likewise if she's kayaking a stretch of whitewater, trying to master sun salutations in a yoga class, posting up in the paint during a basketball game, carving an attractive pattern in a block of striped maple, or shaping a clay pot on the wheel in art class.

But when thoughts are turned inward—for example when there is not much to occupy direct cognitive attention and the brain is "let loose" to roam—the default mode network lights up (by "lights up," I mean that those parts of the brain associated with the default mode network show evidence of blood-oxygen level–dependent signal fluctuations, often referred to as *BOLD* within the neuroimaging clique). When the brain regions associated with our DeMoN are activated, researchers believe individuals may be engaged in future planning—theory-of-mind type thinking, an example of which is thinking in an unfocused way about an anticipated argument with a co-worker and the process whereby one tries to devise rhetorical strategies to ensure victory, reminiscing about the past, or focusing on self-oriented thoughts about moral quandaries.

LET'S BE ANXIOUS TOGETHER!

According to a survey performed by the Association for University and College Counseling Center Directors, anxiety is the number one mental health concern for 41.6 percent of students. The business of treating anxiety is booming—Pfizer, the maker of Xanax, by far the most prescribed pharmaceutical for panic and anxiety disorders, recorded revenues of over $50 billion in 2017.[2] That's right: $50 billion. The fact is, many have been helped by such treatments, but in a scenario all too similar to the opioid epidemic, Xanax has quickly developed into the drug of choice for some populations of young people across the globe, even spawning a musical subculture replete with pop heroes like Lil Xan.

Chronic anxiety can be crippling. Anecdotal experience in the field in the form of interviews with teachers, as well as essays in publications such as the *Chronicle of Higher Education* and others supports this data and then some.[3] Frankly, many students are suffering from various forms of anxiety that span the spectrum from mild stress about taking a new class to debilitating panic. They suffer from all types of mental health issues, but anxiety is a clear winner.

It's heartbreaking to sit across from someone whose stress levels are so high that he breaks into tears discussing the rewrite of an essay, can't make a simple phone call to ask questions about financial aid, deprives himself of vital sleep on a regular basis, and skips meals and class, all because of the nonstop, stress-related hyperdrive mode his brain is in. Others have it worse, self-harming and turning to drugs and alcohol to cope. Some students seem shattered and barely functional. Not all, mind you, but enough that it is a daily presence in the lives of many teachers and educators.

Anxiety is different than stress. At two ends of the spectrum, mild or extreme environmental stressors are no big deal in terms of the way humans have evolved to cope. Primates evolved quite ably to handle a grizzly suddenly popping into view and having to hightail it out of there. Interpersonal relationships and all their related stressors are pretty standard fare, whether one is a low-ranking baboon getting walloped by the alpha male or a student helping a resistant peer study for a math test while also running the social gauntlet of being the captain of varsity volleyball. Environmental stressors can also appear more physical in nature: hunger, cold, physical exertion—humans are supremely adapted to handle those sort of pressures.

But anxiety is different from this type of stress—related in some cases, but with a few marked dissimilarities. A grizzly chase is stressful; the body's fight-or-flight responses kick in, and for a few minutes (until escape or the uglier alternative) everything is in hyperdrive, as it should be: heart racing, mind blank, lungs heaving, adrenal glands juicing up the gross motor system to function at a high level. Anxiety disorder is when there is no grizzly, but an individual is constantly thinking that there *might* be. It's a cognitively based, chronic condition in which negative, cyclical thoughts whirl about, reducing the ability to correctly interpret and perceive the world and its messages.

The problem is that in constantly thinking about grizzlies (or stressing about interpersonal relationships, or success on a test, or chronic worry about how one is perceived by admissions at one's college of choice), the body's stress responses get triggered. Over the course of a bout of long-term anxiety, this is debilitating both because of the cognitive and emotional load but also because being in a state of constant stress wreaks havoc on the body's systems: hypertension, digestive issues, obesity, asthma, premature aging, and sleep disorders result, just to name a few.

If there isn't some measure of control over the DeMoN, anxiety can be the result. If—like many students across the United States—a young person's thoughts turn toward negative, self-eviscerating criticisms during moments when they're not cognitively or physically engaged, then letting the DeMoN run free is a step in the direction of chronic anxiety.

#IMPULSIVEAF

Adolescents and children are naturally impulsive. "Why did you do that?" "What were you thinking?" Most teachers have asked similar questions of their students after a spectacular display of not-thinking-about-consequences. However, impulsiveness is a complicated phenomenon; it's important to remember that a young third grader can be impulsively kind just as well as impulsively thoughtless. In fact, on an evolutionary time line, a bit of impulsiveness can be a good thing, as impulsive decisions could lead to beneficial scenarios. A young hominid takes a wild leap of faith and ditches his family troop, and discovers a new group where he finds a mate. Likewise, a first-year middle school student impulsively decides to take chorus instead of band, and ends up becoming the next Adele.

Impulsivity is a part of normal behavior—just ask any three-year-old. However, developmentally, humans are supposed to "grow out of it" before it mutates into serious behavioral problems: drug and alcohol abuse, risky sexual activities, even crime. It's one thing to snarf down a box of Nilla Wafers while binge-watching Netflix and texting friends and it's another to rob a bank on a whim, but both behaviors are impulsive and can be viewed through both a behavioral lens as well as a neurological one.

Various researchers have defined a link between the DeMoN and impulsive behavior. A group of researchers from Spain, including scientists from the Basque Center for Cognition, Brain, and Language, used parental questionnaires designed to measure the impulsivity of children and then ran magnetic resonance imaging (MRI) scans on the same kids, the idea being that if parents rated their children as highly impulsive, there would be some related brain activity to bear this out.[4] There was.

Children who rated highly in impulsivity showed connectivity between the default mode network and parts of the brain related to action. Or, more succinctly, the part of the brain that makes children do things was connected not to the prefrontal cortex (usually thought of as the master of executive functions in our head, a sort of neurological Emily

Post), but rather showed an increased relationship with the DeMoN, that inward-looking, self-obsessed, and often chaotic part of our brain's systems. Although heredity, upbringing, and social cues all play roles in impulsivity, if a student is toying with the idea of skipping class so they can vape behind the baseball field, it may have something to do with their relationship with their inner DeMoN.

THE DEMON AND THE MADD ONES

A college student—let's call him Brian—had tried to kill himself the day before. His friend—we'll call her Selina—was able to stop him, but in the end it was useless. The next day, after promising her that he was okay and that it wouldn't happen again, Brian hanged himself from some exposed pipes in the bathroom of the dorm. He was a first-year college student at the school where I work.

While I didn't know him personally, some of my students did. They came to me the next day and told me the story in bits and pieces: the desperate and hysterical call from Selina when she found his body, the hopelessness Brian had felt about getting treatment, the slow isolation he created to set the stage for his suicide. Like many severely depressed individuals, the decision to kill himself was probably made long before the actual act. He'd already tried it once that we knew of.

Our campus community was shocked. Naturally, talk exploded on all sides. People were confused, angry, and scared. I advise a student publication that published an essay by a student close to Brian about the death—it angered some who felt it was cavalier, dangerous to students who might be triggered by its content. There were meetings and recriminations. We pulled the piece, but a looming question remained for me: *just what are we supposed to do to help students with severe depression?*

Colleges and universities have worked hard, by and large, to remove the stigma around mental health. But with depression chasing anxiety as the second most prevalent mental illness on campuses around the country, it's clear that understanding the disease—its causes, symptoms, and treatments—is a vital issue in education.

Depression isn't a monolithic disease. It can be a cognitive response to trauma, be genetically inherited, have neurological underpinnings, be spurred by environmental factors, or a combination of any and all of those elements. And although the chronic, at times debilitating effects of depression can have many causes, one associated factor identified by researchers is the correlation between major depressive disorder and the default mode network. That's right: the DeMoN strikes again.

Perhaps it may be helpful to address the relationship between the default mode network and major depressive disorder (MDD). At its core, MDD is inseparable from *rumination*. Rumination is exactly what it sounds like: circular thought patterns of a negative flavor, usually having to do with lack of self-worth or hopelessness. This internally directed focus is self-referential, meaning that those suffering from MDD uncontrollably examine their own feelings and dwell on the negative ones like shame or unworthiness without any hope of resolution. This is markedly different from getting depressed over the number of dead civilians in the Syrian war, for example, which is a situation in which the focus is outward and not self-directed. Simply put: the more individuals ruminate, the more symptoms they exhibit of depression.

Here's where the DeMoN comes in. As discussed, the DeMoN is a self-referential system; it lights up when attention is turned inward. Given that rumination is a major component of depression, it suggests that there is a connection between the default mode network and major depressive disorder.

Before investigating this particular quagmire, perhaps it makes sense to back up a smidge and lay some basic principles that may help explain why there is this correlation between the DeMoN and MDD. One of the more helpful advances neuroscience has made over the past few decades is the ability to map our brains and identify intrinsic functional connectivity (IFC) of the brain. Basically, IFC is different parts of the brain that work together as a system, rather than individual chunks of the brain acting in isolation. Again, we have a scenario where researchers use functional MRI to figure out what parts of the brain respond to various stimuli, activities, and mental states. This is the aforementioned BOLD contrast, which allows researchers to see different parts of the brain and how they're connected to one another. It has been suggested that there are around seven or so of these systems, the default mode network being one. In a nutshell, these large-scale brain maps have provided researchers with a solid means of investigating psychological states—including depression—on a neurological level.

When there is increased connectivity between the default mode network and the subgenual prefrontal cortex, it can indicate a tendency toward rumination, according to research out of the Laureate Institute for Brain Research.[5] This matters because the increased connectivity between the default mode network and subgenual prefrontal cortex can be a recipe for MDD.

The subgenual prefrontal cortex is the newest part of the brain, evolutionarily speaking. It's also responsible for quite a bit of what is known as "executive functions"—social behavior, planning vacations to Disney World, and the way individuals express personality. Guess what else is governed by the prefrontal cortex? Motivation and at least some emotions, as well as what psychologists call "withdrawal behavior," a condition best described by what I do when confronted with a three-hour assessment meeting. Withdrawal behavior is usually a response to unfamiliar people or situations, or even unpleasant occurrences, and is frequently seen in children. However, it is also the type of social withdrawal seen in alcoholics, individuals with damage to the prefrontal cortex, and—you got it—depressives who tend to isolate themselves to avoid social interaction. If a student, for example, is severely depressed, they're probably not all revved up to go to class.

Here is where the rubber hits the road in all this: it seems likely that major depressive disorder is neurologically linked to the default mode network—the DeMoN in the human brain that can accumulate a negative worldview over time. But remember, neural plasticity is real. There are limitations, of course, to how and how much we can rewire our brain.

Many individual who suffer from MDD have other factors, such as early childhood trauma that fuel the feelings of hopelessness typified by depression. But just as brain networks drive behavior, the inverse is also true: *behavior changes brain networks*. What we do—the experiences we have, how we live our lives, whether we're outside and hanging with friends or sitting alone in a room rewatching *The Office* all day—has a direct and measurable impact on the way our brain maps; the intrinsic functional connectivity of our brain's system isn't carved in stone. And as far as the DeMoN in our heads goes, one of the best ways to fight it—and its foot soldiers depression, anxiety, and impulsivity—is to sit and do nothing. On purpose.

For educators to appropriately address the reality of the manner in which the DeMoN affects student well-being, all that is required is a basic understanding of the relationship between unhealthy, ruminative thinking and disembodiment. Teachers should consider whether their students are forced to sit for long periods, are isolated, or are given tasks that allow for the kind of scattered thinking that can be analogous to the way the DeMoN operates; all of these can be precursors to depression and anxiety. Downtime is a good thing—excellent, really, and necessary for mental health—but without providing students with some sense of embodied calm, downtime can actually be the place where negative thoughts creep in. For downtime to really be downtime, for periods of mental relaxation to be relaxing rather than stressful, it is incumbent upon educators to help students develop some sort of mindful, embodied practice to alleviate some of the negative aspects of rumination. Yoga is good, mindful meditation is great. So are dance and mindful walks through the forest or park.

Try this: ask students to spend an entire class noticing their breathing, and saying, to themselves, "calm" on the inhale, and "focus" on the exhale. They must do this while listening to class discussion or lecture, while working through projects. It's in these interstitial moments that students can actively engage their body, simply through focusing on the breath. With practice, this type of mindful breathing can be done just about anywhere: in a crowded cafeteria, on a bike ride to school, walking the hallway to class. Some—many, perhaps—won't actually do it. But if teachers frame the experiment by contextualizing student anxiety, providing evidence to students that these techniques aren't some hippy-dippy ridiculousness but an empirically tested, beneficial activity, those who really need the help may be able to work toward calming the DeMoN in their heads.

NOTES

1. Thich Nhat Hanh, *You Are Here* (Boulder, CO: Shambhala Publications, 2009).
2. Sarah Marsh, "Xanax Misuse: Doctors Warn of 'Emerging Crisis' as UK Sales Rise." *Guardian* (Feb. 5, 2018), https://www.theguardian.com/society/2018/feb/05/xanax-misuse-uk-dark-web-sales-health.
3. Julia Shmalz, "Facing Anxiety," *Chronicle of Higher Education* (Dec. 11, 2017), https://www.chronicle.com/article/Facing-Anxiety/241968.
4. Alberto Inuggi, et al. "Brain Functional Connectivity Changes in Children That Differ in Impulsivity Temperamental Trait." *Advances in Pediatrics*, U.S. National Library of Medicine (2014), https://www.ncbi.nlm.nih.gov/pmc/articles/PMC4018550/.
5. J. Paul Hamilton, et al. "Depressive Rumination, the Default-Mode Network, and the Dark Matter of Clinical Neuroscience." *Biological Psychiatry*, U.S. National Library of Medicine (Aug. 15, 2015), https://www.ncbi.nlm.nih.gov/pubmed/25861700.

SIX
Gandalf of Brooklyn

It has taken decades for the West to embrace ideas of wellness and mindfulness. Asian cultures, particularly those with a healthy strain of Buddhist thinking, have known for eons that separating the mind and the body is a false dichotomy. In fact, the word *namarupa*, derived both from Sanskrit and the Indian language of Pali, represents the union of the body and spirit (or mind) as a whole entity rather than two dueling opposites. Within Buddhist thought, *namarupa* is in a state of constant flux, and the only concept of "self" that exists is dependent on experience rather than abstract concepts. If the conceptual underpinnings of *namarupa* are embraced, the binary mode of Western medicine—body versus mind, psychology versus physiology, health versus illness—comes into question.

In the past twenty years, however, the idea of mindfulness has been, as with any popular idea, commodified, particularly in the United States. Health-food stores now boast mindfulness magazines in their checkout aisles, mindfulness apps haunt the smartphone app stores, and stores and Amazon make piles of money selling everything from incense to meditation cushions to gongs. Many of these manifestations of mindfulness philosophy are Western appropriations of poorly understood Eastern philosophy or commodified placebos for an increasingly digitized and emotionally disconnected elite.

It might seem unnerving that the powers that be in marketing and corporate America have inoculated the public to believe they need $100 leggings from lululemon to do yoga the right way, and affluent schools offer students mindfulness meditation while socioeconomically marginal communities have a "job training" focus in schools. No one would disagree with the importance of jobs for poor communities, but one can still find it frustrating that predominantly wealthy, white students have exposure to wellness programs while minorities and the poor don't. Mindfulness, in many ways, has become yet another privilege found in pricey private schools and educated circles.

And yet the research would strongly support the power of mindfulness meditation and its relationship to cognitive abilities and health, as well as the benefits of embodied approaches to therapy and counseling. Therefore, mindfulness—at its heart an embodied practice—could be an avenue to mitigating student anxiety, increasing academic performance, and alleviating some of the stress that educators feel in modern schools where time to breathe is scarce and the drive to succeed and compete can feel overwhelming.

The movement in education toward embracing ideas of well-being is just beginning, and suggests positive change for the future. In fact, many of the commonplace verbiage used within the mindfulness movement is actually embedded in empirically testable physicality. For instance, the phrase "spiritual health" can be replaced with "interoceptive awareness." The latter can be tested empirically and has a solid body of scientific research supporting it. Even the idea of "chakras" inherent in Hinduism has a correlate within the scientifically verifiable data. The concept of chakras, or bodily energy fields, is deeply intrinsic to esoteric Buddhism and Hinduism. Whether the practitioner of mindfulness buys the ideas of chakras or not, mindfulness meditation has been shown to have positive effects on interoceptive and proprioceptive abilities.

This section deals explicitly with the way in which mindfulness meditation could be leveraged in education to improve student cognitive abilities, overall health, and emotional states.

TAKING NOTE

Jon Kabat-Zinn is one of the most prominent figures in the mindfulness community today. Several decades ago, Kabat-Zinn helped start the Mindfulness Based Stress Reduction clinic at the University of Massachusetts Medical School. The program uses mindfulness meditation to help patients at the UMass Memorial Medical Center deal with chronic pain, stress, and illness. Mindfulness-based stress reduction (MBSR) programs have since spread all over the world, and Kabat-Zinn has gone on to become a prolific author and speaker. Books like *Full Catastrophe Living, Wherever You Go, There You Are,* and *Coming to Our Senses* are accessible and compelling treatises on the value of instituting a regular mindfulness meditation practice.[1]

At its core, mindfulness meditation is based on Buddhist practices. Meditation is deeply concerned with both interoception and proprioception, as the practice of mindfulness requires individuals to become acutely aware of their bodies, both internally, through a cultivation of their interoceptive sensitivity, and externally through proprioception, as mindfulness is best achieved through an awareness of the body's position during meditation. Kabat-Zinn's success at mitigating pain and stress through the MBSR clinic has been studied and verified repeatedly over decades of scholarly interest in the science behind MBSR techniques and practices.

I got a chance to see Kabat-Zinn speak at the Ira Allen Chapel at the University of Vermont. A large, plainly adorned space, the Ira Allen Chapel contained hundreds of fresh-faced undergrads, some crusty looking social revolutionaries, a few buttoned-up professors, and plenty of members of the Vermont elite wearing organic wool pullovers and Patagonia down jackets. The place was surprisingly full, given that Kabat-Zinn was speaking at the lunch hour on a Monday. As he was introduced and began speaking, he gave one the impression of being a Gandalf-like mentor with the voice of a Brooklyn grocer. He's funny and disarming and quite direct—in both his mindfulness CDs and in person he speaks of the "bullshit" we all have to face in trying to find our way to some kind of mindfulness meditation practice. He's likeable.

Kabat-Zinn talked to the assembled crowd about a number of precepts of mindfulness meditation. There are many: for example, non-striving, patience, and acceptance, among others. I listened, rapt, and furiously scribbled notes as he spoke.

To my chagrin, Kabat-Zinn offered to lead us through a meditation but first told us to put down pens and stop "taking notes." "Don't take notes," he said, just "take note" of your surroundings. Well, shit, I thought. Busted. Sheepishly I laid my notebook and pen aside—I glanced around; no one else had been writing stuff down. I was in one of the front rows, and while I couldn't quite see where exactly he was looking, I was pretty sure he was talking to me. I'm always caught up in recording the experience from a utilitarian perspective—how can I use this moment for writing, for teaching—instead of just experiencing moments.

Before he called me out, though, I did get to jot down some interesting ideas, one of which dealt obliquely with embodied cognition and another that offered some insight into the role of the body in learning.

Kabat-Zinn's talk addressed the idea of our "infinite distractibility." He meant that people living in the day-to-day chaos of the twenty-first century are often so consumed with what will happen next, or what is happening elsewhere that they aren't even aware of what's happening at the most local place there is: their own bodies. "Are you even aware of your own body?" he asked us.

It was, admittedly, kind of surprising to realize that in that moment the answer was no, not really. To stop and just *be* with your body is a pretty amazing, clarifying experience. Oddly, it makes me think of those moments in meetings or teaching class when at a particularly inopportune moment my stomach growls and gurgles in frustration—when it happens I'm embarrassed. But why? I have a body. It has guts. Those guts get hungry. Deal with it.

"When we say 'my body,' who's talking?" Kabat-Zinn asked us. The disconnect between the corporeal self and the intellectual, or cognitive, self is so pervasive that it has bisected our language. We distance our *self* from ourselves, and in this sense we mean what we consider to be our conscious, perceiving, thinking selves, from our bodies, the feeling, sweating, farting forms we inhabit. Kabat-Zinn wanted us to recognize that duality and the way it informs our thinking about ourselves, the manner in which we *compartmentalize experience* into bodily and cognitive, when in fact our bodies have been there all the time.

Nowhere is this truer than in education. Funnily enough, although he was speaking at an educational institution (and, presumably, being paid to do so), Kabat-Zinn was critical of the way education in general was approached in the United States. "Education by vaccination" is how he categorized it, suggesting that education conceived merely as a means of "putting things into you" is a flawed approach to school and learning.

Education and learning are active, participatory states, not passive, receptive ones. One doesn't *do* learning; one *is* the learning. It sounds all mystical-crystal, I know. We should be called "human doings" Kabat-Zinn jokes on one of his mindfulness CDs.[2] We are always doing, striving, and moving toward versus being at; always taking notes for the future (guilty as charged) rather than taking note of the present.

How does a shift occur in education such that this inoculative approach gives way to a more grounded, embodied, mindful way of learning? One way to answer this question is to think about how education defines identity, particularly those of students. Education is largely based on a transactional, rather than transformational model. Students are expected to do things: score well on tests, write cogent papers, perform various projects and presentations—all good things. However, educational rhetoric rarely talks about the idea of being, as in working with students on inhabiting a self that learns rather than a self that receives learning. It's a small distinction, but an important one from an embodied learning perspective.

School should function so that kids and young adults can *be* students, not go around acting the part. That means learning for its own sake, not for a grade or a job or a credit hour, but simply because they have cultivated a state of being in which the floodgates of learning are open. Whether they're in accounting or religious history class, their bodies' antennae are primed to receive knowledge; their awareness of their body has finely calibrated their physical perception of experience; their senses are uniquely attuned to decode their environment.

Learning is a state of being, not doing. And while conversations about mindfulness education can get eye rolls so severe from more traditionally oriented academics that they risk pulling their superior rectus eye muscles as they express their insouciance, as anxiety and depression increase to epidemic proportions on campuses all over the United States, it seems a poor choice to ignore the wisdom of a few thousand years of Asian philosophy.

EMBODIED TRAUMA

Ever since Sigmund Freud stroked his goatee and decided that pretty much everything was a repressed sexual desire, psychotherapy has largely been synonymous with *talk* therapy. Countless movies in pop culture support this widespread understanding of how therapy works: a patient sits in a chair (or, if this is taking place in a *New Yorker* cartoon, on a couch) and the therapist sits close by taking notes or staring out the window tossing off observations. Regardless, the key component of therapy and counseling has always been to talk through problems, work on the obstacles encountered in life, and verbally discuss next steps.

There are, of course, many modalities of therapy that embrace notions of mind/body connectivity. These treatments—shown to be effective at dealing with anxiety, trauma, depression, and other mental health issues—can be wide ranging and include traditional practices such as Reiki. One embodied approach to addressing and processing trauma is known as *eye movement desensitization and reprocessing* (EMDR).

EMDR was developed by Francine Shapiro in 1987 as an alternative to more traditional cognitive behavioral therapy (CBT). In CBT, patients work through their trauma or negative experiences by verbalizing them, scaffolding their understanding of the effect those memories have on behavior over time through various dialogue-based means. EMDR, on the other hand, was efficient for some patients according to their own self-reported reflections and through various studies over a shorter period than CBT.

Here's how EMDR works: the client focuses on some aspect of a disturbing memory connected to the trauma they've experienced. Once the patient has immersed herself in the memory, the therapist performs "bilateral stimulation." The most basic form is the therapist holding a finger in front of the patient's face and slowly moving it back and forth. This same back-and-forth action can be accomplished with auditory signaling ping-ponging from right ear to left ear or with a tactile scan during which the patient holds a small object in each hand that lightly, rhythmically vibrates back and forth, left, right, left, right. Regardless of which methodology is employed, the goal is the same: to engage the body in responding to the stimuli to find a location of both physical and psychological trauma.

The terrorist attacks of September 11, 2001, stunned America. Watching planes fly into the Twin Towers that day in 2001 is etched in the memory of many who saw it on television, but for people in and near ground zero it was devastatingly more traumatic. The attacks left many who experienced it firsthand feeling depressed and anxious. Those most affected were individuals actually in or near the Twin Towers when the planes hit.

After the attacks, there was a widespread effort to treat the mental health issues that naturally arose from such a traumatic event. EMDR was used to treat sixty-five people who had been tested for high levels of post-traumatic stress disorder as a result of the terrorist attacks. According to a study of the effectiveness of this particular approach published in the *International Journal of Stress Management*, there was a 50 percent to 61 percent decrease in post-traumatic stress disorder symptoms for individuals who went forward with EMDR treatment.[3]

Although some of the underlying mechanisms of EMDR aren't fully understood, the efficacy of the treatment appears to be undoubtable. The jury is still out on the biochemical interactions, but there are similarities to other forms of therapy that have been thoroughly vetted. Leading up to the sessions during which a client undergoes EMDR, the client completes multiple sessions in which more traditional "talk" therapy is used to establish a framework and history of the adverse memories. It's not unusual for therapists to guide patients through various forms of mindfulness meditation. Body scans, in particular, are used at various points in the process to locate the physical spots that seem to respond to the traumatic memories.

Educators are tasked with many deliverables, and more and more it falls on the plate of teachers to address the trauma students have experienced. To try to teach in the face of trauma is a Sisyphean task; multiple studies as well as common sense strongly suggest that learning under states of intense psychological and physical duress is anemic if not impossible. Therefore, it may be helpful to look at embodied approaches to working through trauma and relate it to education, particularly given the explosive growth of anxiety, depression, and correlated conditions in higher education as well as secondary schools.

It seems highly likely that the success of EMDR is due at least in part to the way that trauma, or negative experiences, are embodied. After all, EMDR is a physical engagement with trauma—a sensorimotor experience rather than a purely cognitive one. In a 2014 article, the founder of the treatment, Francine Shapiro, wrote:

> Except for symptoms caused by organic deficits, toxicity, or injury, the primary foundations of mental health disorders are unprocessed memories of earlier life experiences. It appears that the high level of arousal engendered by distressing life events causes them to be stored in memory with the original emotions, physical sensations, and beliefs.[4]

In short, trauma gets embodied, encoded. When we store memories of traumatic events, we store their physical states as well. EMDR targets the embodied nature of trauma. Intellectually, many individuals suffering from trauma and mental health issues know what is wrong; they are capable of identifying the events in their past that brought about the anxiety or depressive episodes. But *knowing* such a thing, intellectually and abstractly, doesn't necessarily help address the underlying physical embodiment of the trauma. It can take the embodied work of first mindfulness meditation, then embarking on a series of EMDR sessions. It's not as though one rationally processes past trauma through EMDR; rather, it seems to act through a nonverbal, noncognitive, and nonlinguistically articulated process.

It's not only sufferers of major life trauma who can benefit from the embodied approach of EMDR. Shapiro notes, "general life experiences (e.g., relational problems, problems with study or work) can be the source of even more post-traumatic stress symptoms than major trauma."[5]

Stress is an embodied phenomena—it is embedded in a human being's very fibers and systems—and thus EMDR, which in the most general sense shares aspects in common with mindfulness meditation, can be used for many stress-related pathologic conditions.

It's arguable that college campuses are facing an epidemic of mental health issues. Counselors across the nation are reporting more and more students seeking therapy for stress. Counseling centers are overwhelmed with students struggling with depression, anxiety, substance abuse, and social struggles. It's helpful to look not just at the mental health of these students, but also the links to what Shapiro calls "socioemotional behavior problems and cognitive deficits." Students can't learn or study or engage in positive relationships when the very embodied structures that support these acts are compromised by stress and trauma.

Humans store memories in a physical way. This is best exemplified by those moments when a certain smell releases a wave of nostalgia: the scent of the floor wax of your elementary school brings back memories of third grade, or the odor of two-stroke engine exhaust shoots you back in time to a summer landscaping job in high school. The same is true for trauma, but when the traumatic memory is triggered through environmental stimuli because that memory was never physically processed, the individual experiences the trauma of the past as though it's happening right now in the present—sensorimotor perception of the present is informed by unprocessed trauma and stress.

What's deeply informative and supports the notion of mind/body continuity is that it's not *just* emotional trauma. Patients with chronic pain can potentially benefit from EMDR as well. Shapiro notes that individuals who suffer from chronic pain "may actually be debilitated by unprocessed memories encoded with the original somatic perceptions."[6]

Bodies contain the pain of the past. "Unprocessed memory contains the physical sensations experienced at the time of the event," notes Shapiro.[7] Bear in mind that this concept directly aligns with Antonio Damasio's "somatic marker hypothesis" wherein emotional processes—which include trauma, unprocessed memories, socioemotional components, and feelings—dictate behavior.[8] Therefore, the very physical reality of embodied trauma orchestrates emotions, which in part dictate behavior. It looks like this:

Trauma and unprocessed memory > emotions and feelings > behavior

The issue for students who've been exposed to trauma or negative stimuli throughout their development is that research has shown that untreated trauma or unprocessed memories of adverse treatment such as abuse lead to a host of other issues including substance abuse, obesity, severe depression, and anxiety. There's a physical correlation as well. Individuals who've experienced childhood trauma are more likely to suffer from diseases of the heart, lung, and liver, and are more susceptible to certain forms of cancer.

What's an educator to do? With the rate of self-reported anxiety and depression increasing on college campuses as well as earlier in childhood, and with many teachers instructing children who suffer from the stress of broken homes and traumatic environments, it can feel overwhelming. After all, educators are trained to teach quadratic equations or explain Plato's allegory of the cave; the teacher who also happens to be a trained therapist is rare. But knowledge can help teachers as they navigate these increasingly troubled waters. To regroup:

1. Trauma and bad memories are embodied.
2. Although CBT is an effective treatment for individuals suffering from adverse experiences in the past, so is EMDR.
3. EMDR operates on the same general principles as mindfulness meditation. Namely, that the process of recalibrating psychic states begins with recalibrating the body.
4. It may be helpful to think about students suffering *physically* from *emotional* trauma.

Many educators have expressed feeling overwhelmed by the sheer number of students who confess or exhibit signs of anxiety, depression, and self-harm—the walking wounded are many. EMDR is one example of the power of using an embodied approach to learning. Recognizing that learning and healing happen through the body is a vital component of developing a teaching philosophy that takes into account the embodied nature of student experiences. If students both experience developmental milestones and are hurt through their sensorimotor experiences in the world, then perhaps as a teacher it may be a good idea to incorporate a bit of mindful, embodied practice into the classroom.

It's 9:30 a.m. during midterms. My students shuffle into the room, puffy faced and yawning. When I ask, "How is everyone?" I'm met with the same answer nine times out of ten: "tired and stressed." So I dim the lights and lead them through a simple body scan meditation. We breathe, then focus on the toes, feet, ankles, calves, on up through the body. With each exhale, I ask them to release the tension from whichever body part we're focusing on. We end with a series of breaths where on the in-breath the students say the word "strong" to themselves and imagine themselves firmly rooted to the earth like a mighty oak tree, and on the out-breath they say "calm" to themselves and let their shoulders relax and melt like butter. A few breaths to come back into the room and we're done. Six, seven minutes, tops.

I'm not a Zen monk with decades of meditation experience; I doubt many Zen monks binge watch *Stranger Things* on Netflix while snarfing down chips and gulping Vermont microbrews. But with a fifteen-week semester, I spend a total of almost forty hours with each group of students. That's a long time. I'm confronted with their struggles every day. If I didn't try, in some small way, to acknowledge the embodied nature of learning, the physical correlates of learning, and the way sensorimotor perception informs a sense of "self," then I'd be doing them a disservice.

The students shake themselves back to reality after the brief mindfulness exercise. One of my students, a twenty-eight-year-old who's returned to college after eight years in the Navy, asks me, "Professor Shonstrom, do you meditate regularly?" I could lie, and provide a (false) example for him. Tell him I wake early, say at 4:30 a.m., meditate for an hour, and then make my son's lunch while emitting loving-kindness vibes. But I don't lie.

"I try to," I say. "But mostly I fail."
He chuckles. "You try?" he says.
I shrug. It's all you can do.

NOTES

1. Jon Kabat-Zinn, *Full Catastrophe Living: Using the Wisdom of Your Body and Mind to Face Stress, Pain, and Illness* (New York: Bantam, 2013); Jon Kabat-Zinn, *Wherever You Go, There You Are: Mindfulness Meditation in Everyday Life* (New York: Hyperion, 2005); Jon Kabat-Zinn, *Coming to Our Senses: Healing Ourselves and the World through Mindfulness* (New York: Hyperion, 2005).

2. Jon Kabat-Zinn, *Mindfulness for Beginners: Reclaiming the Present Moment and Your Life* (Sounds True, 2012), CD.

3. Sandra A. Wilson, et al. "Stress Management with Law Enforcement Personnel: A Controlled Outcome Study of EMDR Versus a Traditional Stress Management Program." (New York: Kluwer Academic Publishers-Plenum Publishers, July 2001), https://link.springer.com/article/10.1023/A:1011366408693, 179–80.

4. Francine Shapiro, "The Role of Eye Movement Desensitization and Reprocessing (EMDR) Therapy in Medicine: Addressing the Psychological and Physical Symptoms Stemming from Adverse Life Experiences," *The Permanente Journal* (2014), https://www.ncbi.nlm.nih.gov/pmc/articles/PMC3951033/.

5. Shapiro, "The Role of Eye Movement Desensitization and Reprocessing (EMDR) Therapy in Medicine."

6. Shapiro, "The Role of Eye Movement Desensitization and Reprocessing (EMDR) Therapy in Medicine."

7. Shapiro, "The Role of Eye Movement Desensitization and Reprocessing (EMDR) Therapy in Medicine."

8. Antonio Damasio, "The Somatic Marker Hypothesis and the Possible Functions of the Prefrontal Cortex." *The Prefrontal Cortex Executive and Cognitive Functions* (Apr. 1998), doi:10.1093/acprof:oso/9780198524410.003.0004, pp. 36–50.

SEVEN
Bodies of Literature

The act of reading is fantastically complex. When viewed through a particular cultural lens, it is the symbol of conscientious study and diligent scholasticism; within a different framework, it can be seen as a radically subversive act. One thinks of the uniformed school boy industriously bent over his books contrasted with the wild radical, red-rimmed eyes gleaning revolutionary propaganda from samizdat pamphlets.

This Jekyll-and-Hyde duality of reading can been seen perhaps most clearly in school libraries, where books are assigned and read and catalogued and also, sometimes, banned. Reading is both the foundational core of education—five- and six-year-olds plugging along through Richard Scarry and Dr. Seuss—and the apex of learners' intellectual lives when they attempt to parse dense texts by Foucault and Derrida, scale the epic heights of the *Mahabharata,* and try to contextualize and reflect on the multitudinously meaningful essays of James Baldwin.

Historically, reading can be the most solitary, private act imaginable, or the shared experience of millions. Some turn to reading as an act of withdrawal—a deliberate closing off of the outer self in favor of a completely internal experience—while others use books as a conduit to rich friendships and shared experiences through book clubs, author readings, and, in large part, school itself.

But for all of reading's contradictory profiles, perhaps none is more revelatory than the way reading has been defined for centuries as a process of imagination and intellect, an ethereal act in which the reader absorbs the texts and creates immaterial pictures in their minds. It is a purely rational and abstract transaction between the words on the page and the reader's mind, a vapor of cognitive action. And although reading can be seen in this light—as you read these words, this is how, in fact, it may feel—the reality is strikingly the opposite.

Reading is the most embodied learning experience there is.

GLIBLABBERFLUXAMATE

To begin to understand the implications of this profound reality—*reading is embodied*—and what it means for education, it is necessary to position reading within related cognitive and environmental scenarios. First, reading is a result of experience. If you teach a child to read the word "dog," than what is implicit is that the child has an experience of "dogness" to make sense of "dog." Now, they may not have seen a real dog, but they understand at some level that there are animals, pets, and other creatures, and can therefore position the word "dog" within a context.

For instance, if I ask a student to read and memorize the spelling of "milfoil" without the student knowing what "milfoil" means, the word will essentially be meaningless. Try to remember the word "gliblabberfluxumate" two days from now—likely you won't be able to, because you don't know the word and can't contextualize its meaning (and also because it's not a real word).

If, on the other hand, one tells a student that "milfoil" is a submersed, aquatic freshwater plant that originated in Australia but is now a persistent and pesky invasive species in rivers and lakes of North America, the student will be able to draw upon their knowledge of lily pads or seaweed or other experiences of plants that grow in water and more or less understand what "milfoil" means. Their experience of other plants has injected a previously unknown word with meaning—an assemblage of associations that now cements "milfoil" in a particular place, thus making it easier to remember, recall, and reuse.

The second way in which reading must be considered is to realize that it is an act, a physical event, an experience had by the body. You're having it **right now**. As you read R-I-G-H-T N-O-W it is an embodied act, not a solely cognitive one. Researcher Anne Mangen puts it thus: "Reading and writing are multisensory, cognitive processes relying on complex perceptual sensorimotor circuits and connections."[1]

There is a connection between direct experience, reading, and language processing. Put another way, reading exists as the fulcrum between the comprehension of tangible, physical reality and conjectural or abstract thought—but even abstract thought is not some disembodied conceptual realm, but rather is founded on, and shot through with, the understanding borne of the reader's physical presence in the world. A researcher named Michiel van Elk from the Netherlands describes reading as the way in which we assess and act: "language comprehension can be described as procedural knowledge—knowledge how, not knowledge that—that enables us to interact with others in a shared physical world."[2]

Again, this is where direct experience comes into play. Without a bankroll of experiences to draw upon, it is possible that reading comprehension will be affected negatively. Life well lived begets fluidity in text comprehension.

One way to think about the embodiment of language is through the old story mentioned earlier of the Inuit who have no words for trees, as they live in a treeless, permafrost land, and yet have many words for snow. Their language is based on their experience.

OLD-SCHOOL LEGOS

The reason for this connection between the body and reading comprehension comes down to the fact that as biological entities, humans are built with old-school, interchangeable Legos. Human beings don't necessarily have fancy specialized equipment—they have to work with the same systems they crawled out of the muck with about 530 million years ago. In other words, humans have a basic set of building blocks and physical, musculoskeletal, and neurological structures that they have to work with. As a species, humans don't have specialized equipment to develop new behaviors—they have the same basic stuff that other related species have evolved to cope with the environment. Thus, humans need to use the tools of the body to shape the work of the mind.

Bodies are built on the same basic blueprint as most other complex organisms, with only so many structures and systems. Over the course of evolutionary development, different parts and systems of the body are recruited for new skill sets. As a species humans adapted the physical ability to sense their surroundings, to be aware of the body in space and time—to perceive and act according to the dynamic and subtle environmental stimuli presented. See a saber-toothed tiger leap from behind a bush? Run. Smell delicious, ripening mangoes? Follow the scent and eat. Hear trickling water, feel the mist cooling the skin? Go have a drink and a dip. This interconnected play of smell, sight, touch, and the resultant physical response is often referred to as "sensorimotor experience."

The connection with reading is best summed up through the experience of reading words with strong associations for the senses. For instance, reading words like "onion" and "vanilla" actually engage parts of the neurological systems normally involved in perception—in other words, when you just read the word "onion," as far as your body's sensorimotor system is concerned, you smelled it. The same part of the brain lights up whether we take the words off the page with our eyes or the pungent odor through our nose.

In other words, just like other theories of embodied cognition having to do with behavior, decision making, and writing (just to name a few—the list keeps growing), reading as a sensorimotor experience rests on the same basic principle: the reading and processing of language automatically engages the same biological and neurological pathways and structures that are used during perception and response—otherwise known as *experience*.

Consider a final example. When reading a statement like "open the door," the same neural architecture that is used to actually open a physical door is fired up and activated—thus, the more we've lived, the more direct and varied experiences we've had, the more easily we can recruit our neural systems (our entire bodies, really) in reading comprehension.

The ramifications for the classroom are profound. A tacit understanding of the embodied nature of reading leads us to understandably question a number of ongoing educational approaches that have crept into the mainstream; namely, reading on tablets and laptops and using the web as the primary means of building literacy skills.

THE SCREEN IS MORE MEAN THAN IT SEEMS

Reading is a physical act, and therefore the substrate or surface that contains the text and is read, held, manipulated, and carried about *matters*. Within the United States, the prevalence of the screen as the primary means of transmitting information has become almost unquestionable. My son—who's entering sixth grade—has his "own" Chromebook in the classroom. It was used to write stories, play math games, and download screen savers of Legolas piercing orcs with arrows. What has become clear is that tablets and laptops are replacing older technologies such as books and paper. Not entirely—the libraries of our schools are still bustling hives of reading—but the encroachment of screen-based pedagogies is undeniable.

What does education lose when the process of developing literacy skills is transferred from the page to the screen? Research is beginning to demonstrate that the most crucial aspects of reading are affected. Because reading is a physical act, the relationship between the reader and the thing being read isn't secondary to the experience.

Books must first of all be held. Most screens can be read without necessarily touching the laptop of iPad or Kindle (although most e-readers are held in some way). With books, the reader must come into direct, physical contact with the text. A reader has to pick it up, flip the pages, angle it to get better light, notice the cover, locate the spot where he or she left off, and even mark up the margins with notes.

Normally, the argument by the old, musty, gray-haired contingent goes something along the lines of preserving this tradition, maintaining contact with the historically relevant aspects of reading through reading books. Some critics will even argue that it's the reader's civic duty to buy books to prop up the publishing industry. These are all decent arguments, but the benefits of reading books are actually much deeper and more profound.

There's an essential lineage of reading that we have to understand to fully appreciate the importance of books versus e-readers.

- Language is metaphoric.
- Metaphors are largely embodied.
- Reading comprehension is based largely on our relationship with our own body.
- Our relationship with our body is based on experiences.

EMBODIED METAPHOR

Language is largely symbolic, in both written and verbal forms. With the exception of lovely onomatopoetic words like "bang" and "sploosh," most words have little to do with their meaning in terms of sound. Nor do words look like the thing they're describing.

If one writes the word "sneaker," it looks no more like a Nike cross-trainer than any other word. So at its most basic point, to understand language on any level is to understand its symbolic nature, where one thing is representing something else.

The most highly evolved form of this linguistic bait and switch is, of course, metaphor, which could be argued to be the most basic meaning-making part of language. Metaphoric language has been around since the Mesopotamians first started using a cuneiform stylus to record livestock transactions as well as the deeds of kings. We use metaphor (or its sibling, simile) all the time to describe the world around us as well as our own experiences. "This toast is as hard as a rock!" "I'm on fire today!" "Reading Shonstrom's prose is as uncomfortable as wading through a thousand gallons of melted cheese!" Although this seems self-evident, when we begin to consider the embodied nature of writing, this largely metaphoric approach to understanding the world around us takes on new weight.

As the groundbreaking work of Mark Johnson and George Lakoff has demonstrated, metaphor is largely embodied.[3] Embedded within most metaphors is an inherent reference to our bodies—to the way in which our bodies relate to themselves and our environment. Essentially, readers use embodied experience to make sense of their lives. The most basic of these embodied metaphors come from early in life, where humans begin to equate warmth and closeness (think of a parent cuddling an infant) with positive attributes. Thus, a "close" friend is one who can be trusted, and feeling the "warmth" of friendship is a good thing, not a sign of impending flame. Even when discussing trials and tribulations, common language employs a bodily metaphor based in carrying burdens, as in being "weighed" down by troubles. We describe abstract concepts in the negative by using what are clearly negative embodied experiences; computers "crash" and the economy "took a dive" this week.

Given that the very primacy of language is metaphor, and metaphor has been argued by Lakoff and Johnson to be largely a result of our relationship with the body, it follows that understanding bodies must be connected to understanding language—and this turns out to be true.

EXPERIENTIAL LANGUAGE COMPREHENSION 101

Reading comprehension has been shown to be positively increased when incorporated with physical experiences. Again, if you've swum in a lake and all of a sudden had the slimy tendrils of some lake weed wrap around your legs, and someone introduces the concept of the invasive "milfoil" species to you, your comprehension level will be higher than someone who's never taken a plunge into the cold waters of the lake. But even beyond that, if, through sports or any movement-oriented activity, a student has developed her or her proprioceptive and interoceptive senses, and has a finely calibrated relationship with their own physicality, the embodied metaphors of Lakoff and Johnson will ring more deeply, truly, and clear. When traffic is "congested" and a flight is "running" late, those individuals with an innate, corporeal sense of language as it inhabits the body will intuit nuances of meaning that someone without a wide breadth of experience won't.

This relationship between experience, metaphor, and language comprehension argues strongly in favor of an experiential-based curriculum not just for kindergarteners, but all of us, regardless of age. Let's reduce it to a very simple and unscientific equation just to wrap our heads around what all this means:

Embodied Experience + Embodied Metaphor = Reading Comprehension

This brings us back to your favorite e-reader, Kindle, or iPad. It matters what students read *from* not just because we are looking backward in stodgy bitterness at better days, but because reading is a physical act, thus it must be enacted physically. Put another way, everything about reading is based in the concrete, tactile experience of being an actual, physical human being in the world.

Screen-based text interfaces limit the corporeal, haptic relationship of reading. Flipping pages in a book reinforces the narrative of the book itself. The electronic interface doesn't allow for the same interaction with the arc of the story—the trajectory of experience related both to the content of the book as well as the experience reading it. In fact, it could be argued that learning itself is a narrative, and that by undermining the primacy of the book, which encapsulates the metaphoric meaning of the developmental arc in a way that e-readers just can't, digital interfaces used to foster literacy and reading comprehension aren't simply a benign option—they're reducing the potential impact of education on the student's sense of self.

THE SWEETNESS OF WORDS

When young Jewish children begin to attend *cheder*, or school, at around the age of three, they also begin to learn to recognize letters of the alphabet. Rabbis will often dab a little honey on letters of the alphabet so that the young scholars begin to associate the Torah with a sweet taste—the words of Yahweh and the Torah affixing themselves as delightful in the child's mind.

This act—the physical ingesting of words—stands as an excellent introduction to how to incorporate the science of embodied cognition into a language comprehension curriculum. Licking honey off the page is a wonderful reminder of corporeal nature of reading—reading and language comprehension isn't simply aided by movement or work with the body; it lives there, is incorporated into our very physical selves. One of the clearest ways to capitalize on this in the classroom is to have students stand and read aloud.

Remember speech class? I was instructed to stand still, keep my hands at my sides, not fidget, and deliver three minutes of microwaved speech. Thank goodness handheld technology hadn't progressed to the point that I was filmed in those moments—it must've been like watching an animated cadaver talking.

Now, as a teacher, I get paid to talk, and I very rarely stand still. In fact, if you dial up a good TED Talk on your computer, one of the things you'll notice is how much people move—people giving a really good talk that's engaging and interesting and entertaining move about. They stride across the stage, gesticulate, point, shuffle—they move their bodies. Not all, mind you; there are the few who can stand, fence post–like, and hold us in sway with their penetrating gaze, but most of the dynamic speakers I've watched move around. A lot.

This makes us wonder: why do we have "silent" reading times, when kids are asked to sit at desks? Why not read aloud, while walking around? My son reads all the time, and often while walking, laying on the ground, swinging his legs; he's in constant movement. Watching someone enraptured in the text can often be accompanied by embodied actions that seem automatic: twirling hair, chewing of lips, tapping foot, tongues sticking out. The rabbis had it right—reading is an ingestion of the text. Reading is nothing short of taking the essence and meaning of the words into our bodies.

Reading out loud has a number of benefits. When children are asked to read aloud, it increases comprehension. This is in part because silent reading—or just the text itself—is rather flat in affect. Punctuation serves as a kind of musical notation—a way to further identify meaning, emotion, and emphasis. Really, the English language is constructed as a spoken medium. When children are able to read aloud, this embodied approach to language increases comprehension but also teases out nuance. In their recent book on the power of reading with children, *Read with Me*, Susan Cleaver and Munro Richardson point out the importance of reading aloud to young children and add that there is a difference between "reading *to* a child and reading *with* them."[4] Reading aloud to one another is an incredible transference of tone, mood, context, and meaning. I still read to my college students, and the effect is the same every time. Comprehension jumps, questions abound, and interest peaks. This is partly because reading aloud taps into the central tenets of our narrative selves: we are storyborn creatures to be sure. Take any group of elementary students, and intone, "Once upon a time, in a dark wood . . ." and you'll have their attention. We're built to hear stories, and how the stories are told, and the way in which we embody their meaning and context, makes all the difference.

Here's an example, using one of my favorite jokes told to me by the same Jewish friend who related tales of licking honey off the Torah. A young man and his Jewish grandfather head to the beach one day. As they approach, they see a sign posted. The young man stops, but the old man heads right past the sign and down onto the beach, where he proceeds to take off his shoes, strip down to his swimsuit, and wade out into the water. After his dip, he comes out of the water, and the young man says to him, "Didn't you read the sign?"

"I did," says the Jewish grandfather.

"Well, it said the beach was closed," said the young man. They walked back to the sign, which read:

Beach Closed
No Swimming Allowed

"See?" says the young man.

The grandfather looks at him, and says "what that sign says is "Beach Closed? No! Swimming Allowed.""

It's funnier when my friend says it with his grandfather's thick Brooklyn accent, but you get the idea; intonation and what is referred to as *reading prosody* matter. Language is closely related to music, and in fact when we read aloud well, the same parts of the brain that light up with music-related tasks also go into overdrive—another way in which reading is embodied. If the same parts of the brain fire when you play piano as when you read aloud, it suggests a deep level of interconnectedness between the way we move our bodies, whether it's our fingers or our mouths, when engaging in communication.

What this means for children learning to read is most clearly articulated by a researcher with the wonderful name of Paula Schwanenflugel (itself a fun word to read aloud) who has noted in various research projects that reading comprehension goes hand in hand with reading prosody.[5] Children who are first learning to read do so with poor prosody—just getting the words right is enough. However, as reading comprehension gets better, children naturally develop the type of performative aspects of reading we associate with good out-loud readers such as using higher tones at the end of questions—"Are you going to the park?"—and flat, decreased tones at the end of sentences such as, "No, my bike is busted."

Not only is comprehension improved through embodying text by reading aloud, but the ability to remember what we've read is improved as well. When we read silently, we are using our visual pathways to remember the words. We remember the sentences and stories and information because it is something we saw.

For instance, if you read the words "sticky spider bicycles" you'll remember it partly because of the associations you have with the words, but also because you've actually looked at the three words and created a visual memory of it. This is exemplified by individuals with photographic memories—they have more robust visual pathways relating to information than those of us who can't remember the main character of the novel we're reading. However, by reading "sticky spider bicycles" out loud, we've embodied comprehension through the use of our mouths and lips, thus creating a muscle memory of the phrase, and also have created an auditory pathway to the memory. This process is called the "production effect."

Research detailed by Colin MacLeod, Nigel Gopie, Kathleen Hourihan, Karen Neary, and Jason Ozubko in the May 2010 issue of the *Journal of Experimental Psychology: Learning, Memory, and Cognition* supports the notion that reading aloud improves memory.[6] But why it improves it is fascinating; for instance, we have a tendency to notice what is distinctive about a list of names better when we read it aloud. When reading silently, "sticky spider bicycles" is just a random assortment of words, but read aloud, its oddballness becomes a prevalent characteristic. It's rather poetic construction (if I do say so myself) is only apparent once read aloud; the alliteration of the first two words, the rhythm of the syllables, the internal rhyme of the *i* in "spider" and "bicycles" only come to life, and get locked into our brains, when read aloud.

Reading aloud embodies both comprehension and memory—we talk about "muscle memory" when performing physical tasks, but the reality is that embodied cognition supports the idea that *all* memory is muscle memory—including reading.

Reading is not the same as reverie or daydreaming. They are closely related, sure, and reading certainly can lead to both those states. But reading is a highly engaged, bodily experience. Next time you're reading a thriller late at night, and have arrived at the crux of the mystery, note how your body is tense, how your breathing is shallow, maybe your eyes are even a bit wider than normal. You're having an embodied reaction, a sustained complex physical interaction with the experience of reading.

Teachers who read aloud with children, engage with the text playfully, and create a dynamic, two-way experience with books will have students who profit in terms of reading comprehension. In addition, teachers who default to real books with pages instead of e-readers allow for the haptic interaction of reader and text; children physically move through the book as they move through the story. Progress is marked and made and noted. It's a journey lost in the flat abyss of digital readers. And finally, reading comprehension is based in large part on experience. The more teachers can get students out in the world, embodying their learning, the more books will come alive in the classroom.

NOTES

1. Anne Mangen, Jean-Luc Velay. "Digitizing Literacy: Reflections on the Haptics of Writing." *IntechOpen* (Apr. 1, 2010), https://www.intechopen.com/books/advances-in-haptics/digitizing-literacy-reflections-on-the-haptics-of-writing.

2. Michiel van Elk, "Embodied Language Comprehension Requires an Enactivist Paradigm of Cognition." *Frontiers in Psychology* (Dec. 27, 2010), https://www.ncbi.nlm.nih.gov/pmc/articles/PMC3153838/.

3. George Lakoff and Mark Johnson, *Philosophy in the Flesh: The Embodied Mind and Its Challenge to Western Thought* (New York: Basic, 1999).

4. Susan Cleaver and Munro Richardson, *Read with Me: Engaging Your Child in Active Reading* (Lanham, MD: Rowman & Littlefield, 2019), xvii.

5. R. Benjamin and Paula Schwanenflugel, "Text Complexity and Oral Reading Prosody in Young Readers." *Reading Research Quarterly*, 45(4) (2010), 388–404.

6. Colin MacLeod, Nigel Gopie, Kathleen Hourihan, Karen Neary, and Jason Ozubko, "The Production Effect: Delineation of a Phenomenon," *Journal of Experimental Psychology: Learning, Memory, and Cognition* (May 2010). Available at: https://www.ncbi.nlm.nih.gov/pubmed/20438265.

EIGHT
Prose in Repose

Writing is an act of magic. At the same time mundane and transcendent, the act of putting words on paper has within it something of the occult. Prior to the actual finished, written piece, what even *is* an idea? Random snippets of line and sentences; lumpy, undistilled memory; a feeling more than some well-defined concept or theory. Sometimes it's not even any of those things, but rather a faceless urgency. So pencil gets drawn across the surface of the paper, and the hope that something will come of it burns brightly.

Of course, there are also times when a cogent thought develops for students and they are able to write it down in its entirety, complete. Those moments are rare. It's possible there are writers who have the ability to string together whole sentences, paragraphs, pages even within their skulls, then sit down and fluidly pour forth line after line of clear ideas. It often seems that some writers—one thinks here of Virginia Woolf—have a highly evolved ability to do exactly that; it's as though Woolf is walking about the streets of London, stringing together endless, resonant sentences like the rows of a knitted scarf, one yarn binding all together and then placed, as whole cloth, on the page.

Writing is an embodied practice, more like dance, perhaps, than other forms of creation. In an excellent essay, "Dance Lessons for Writers," Zadie Smith makes the connection resolute between the two forms.[1] Writing is not only like dance, and writers are not only like certain dancers, but they seem, in some ways, simply different versions of the same expressive art.

Prior to the movement of pencil on page, the story is inert, humming with possibility, maybe, but without movement. And as a dancer waits her turn in the wings, she prepares to bring a narrative to life through the movements of limb and torso, head and foot, just as the writer uses the movement of the pencil to etch life out of nothingness. To bring feeling and thought into existence through embodied movement.

Perhaps it's a bit of a poetic stretch. After all, both writing and dancing have hours, days, years of preparation and rehearsal, drafting, and failure. And frankly, the process of writing is, more often than not, monotonous, boring, frustrating, a time-suck, and rather disappointing. Beckett may have been onto something when he urged, "Fail again. Fail better."

But even without cherubic trumpets blowing every time an individual sits down to write, the act can be a little mysterious. One of the reasons writers write about writing and creativity (Stephen King's *On Writing*[2] or Elizabeth Gilbert's *Big Magic*,[3] for example) is that it's such a nebulous enterprise open to interpretation; it is inscrutable and elusive and hard to pin down. But one thing is for sure: it takes a physical act to bring it to life.

A writer—be it an eighth grader struggling to churn out a book report or Virginia Woolf herself—has to physically put it down somewhere, and conjure thought and line and idea out of thin air onto the page. It is an embodied process, and interplay between sensorimotor function, perception, cognitive gear-spinning, fine motor control, and even gross motor movement. For some it can also involve ridiculous amounts of coffee.

I hold a pencil in my hand. I use the classic pincer grip, thumb and forefinger arched and tweezing right above the business end, resting where the yellow gives way to conically shaved wood that tapers to a fine point of graphite. My middle finger is the third flying buttress supporting the pencil, but it's curled under a bit and the pencil rests on the side of my final knuckle. For those of you interested in such nerdy ephemera, it's called the *dynamic quadrupod grip* by occupational therapists.

Usually, I'm at a table of some kind with feet on the floor, head canted forward toward the page. Writing these lines now, for the first draft of this chapter, I am sitting on a couch in our den, slumped against the armrest, legs stretched out in front of me on the cushions, tired already at 7:55 a.m. after a morning of lunch packing, oatmeal stirring, egg scrambling, bus waiting, and the general cacophony and chaos of the average suburban work and school day.

When I write, I hear the scratch of the pencil graphite against the paper. It's not entirely dissimilar to the way telegraphs sound in the movies; there's a rapid-fire series of scratches punctuated by periodic breaks when I mentally need to search for a word or complete a thought. "Erik is trying to write. STOP. It is going exceedingly slow. STOP. Would rather binge watch Bachelor in Paradise. STOP."

My handwriting is just this side of atrocious. I can, when I need to, write in what I refer to as my third-grade-teacher-blackboard-writing: clear and straight and even. Now, however, I am trying to get some writing done before a 9:30 a.m. class, so my penmanship is hurried and bears more resemblance to chicken scratches in the dust than anything. My lines are crooked and, as usual, my writing starts out small and dark but as I tire or hurry even more to finish and the graphite dulls from a needle point to a snub-nosed, blunted tool, the actual letters get larger and there's more space between lines and words. It's all a bit of a mess, really, not helped by the fact that I'm jotting this all down on the backside of a research article on yogic breathing that I printed out but haven't read yet.

But it's all very human, and my own process is similar in many ways to the process educators attempt to instill in students throughout their school years. It is, in multiple ways, a deeply recursive, embodied experience. Students must think about what they're writing while they're writing it. The whole enterprise hinges on the haptic experience of moving the pen across the page, calling to mind not just the shape of letters but their ordering—digging into both the memory and analytic capacity when trying to spell the word *atrocious* while not fully awake.

Sitting next to me—mute and terrible—on a coffee table is a laptop computer, the same technology that students in the United States are primed to use through typing tests, digital projects, and the full immersion in technological education. educational technology has become so prevalent in K–12 education as well as college that it has seamlessly worked its way into every aspect of education.

Most schools now operate on a "learning management system" (LMS) that allows teachers and professors to upload assignments and readings as well as collect student work digitally. Many of the LMS platforms offer chat and discussion board functions, video feedback tools, grade books, and attendance. A student can see, in real time, exactly what their grade is and how many assignments they've either done or left incomplete. Some students will state that this digital organization is helpful; others will say that it provokes constant anxiety. What is clear is that there is a large-scale, nigh-undeniable cultural expectation that the work of writing that was once one of the central tenets of education—one of the three Rs—has now become almost entirely virtual and electronic.

It's in this way that the essential embodied nature of writing is under siege from technological alternatives. Although some technological devices may be helpful for students, others may be changing the way in which they interact with the written word, and it's entirely believable that at some points the digital medium is detrimental to what educators are trying to do in terms of the personal and intellectual development of their students.

THE STUDENT'S *MISE EN PLACE*

Writing—in its preinternet, analog form of paper and pens and pencils—can feel more in line with slinging hash for the breakfast rush than some detached cerebral task. Just as making bread is both the most basic culinary practice one can imagine, yet imbued with the startling alchemical magic of fermentation, writing can seem to be a pretty straightforward enterprise of putting words on paper, and yet it too has a transformative element to it.

Cooks talk about their *mise en place*, which is a French phrase that refers to the physical setup of their particular station. Having the right ingredients and tools nearby, exactly where you need them to be, is vital if you're going to cook well. Cooks and chefs will talk about "being in the weeds," an expression used to denote an intensely busy shift. During the chaos of a full house, the last thing a cook needs to be doing is searching for a pair of tongs or a spatula.

Cooks can be fanatically obsessive about their "meez" and attached to particular cutting boards, slotted spoons, and frying pans to the point of fetish. For a humorous take on this, and on how cooking is truly an embodied experience, Anthony Bourdain's *Kitchen Confidential* explicates the physical, arduous nature of cooking. In many ways, writing is similar to this process of developing the correct tools and physical space to do it well.

One of the possible dangers of digital mediums is that they reduce, or remove, the physical elements that necessitate a corporeal engagement with the process of writing. Tapping keys is not the same as writing longhand, just as punching buttons on a microwave is not the same as testing whether or not a steak is done by poking it with a finger. What lies at the heart of both cooking and writing (and many other physical activities) is the relationship of the body to tools, and the way that tools can shape the activities and behavior of a body.

Evolutionary biologists are forever trying to pin down what it is about our species' ancestry that gave rise to our current status as the most technologically advanced, complex species on the planet. Big brains, fire, social coordination, tool use, persistence hunting, opposable thumbs—all have been touted as the thing that gave a one-hundred-pound bipedal hominid the chutzpah to colonize and change the planet. It's an argument that's rich and helpful to move the ball down the field of understanding human behavior; it can even be fun to follow the developments if you like looking for snarky jabs in peer-reviewed articles aimed at other academics.

The outlines of this book are too far outside of those disciplines to join that particular academic scrum, but it seems clear that writing may be an extension of some of the earliest evidence of embodied cognition, particularly in the way in which tool use and the development of our neurochemistry seemed to have developed in conjunction with one another. At the very least, the relationship between manipulating our environment through physical action, the reception of sensorimotor information, and cognition bear some scrutiny.

TO A PERSON WITH A HAMMER, EVERYTHING LOOKS LIKE A NAIL

Tool use—knapping stones to make spearpoints, using bone awls to punch holes in leather, carving arrow shafts—has been linked persuasively to cognitive development, learning, and potentially language development. One of the most well researched accounts of the way the work of the hand led to cognitive development is the book *The Hand* by Frank Wilson.[4] Wilson contends that it was humans' ability to manipulate the world around them by hand that gave rise to cognitive processes that in turn opened the pathways to the complicated societies we have now. In this sense, learning was very much embodied, as the hands quite literally grasped concepts (pun intended) and passed that information along to the brain.

The way we physically interact with the world is directly connected to the way in which we cognitively process experience. Or, put another way, experience isn't some abstract, mental picture or series of imagined events, but rather is our body's interaction with our environment. The interface between our body and the tools we use *matters*. As the old adage suggests, everything looks like a nail to a person with a hammer.

The shift that's currently well underway from page-based culture to a screen-based one can't be overstated. Despite the book's relative newness—Gutenberg invented his printing press in the mid-1400s—the book rapidly became the repository for knowledge within the literate world. That is indisputably changing, with more text and information available on the web every day.

Semiotician Gunther Kress—a scholar researching digital literacy—puts it this way:

> The combined effects on writing of the dominance of the mode of image and the medium of screen will produce deep changes in the forms and functions of writing. This in turn will have profound effects on the human cognitive/affective, cultural and bodily engagement with the world, and on forms and shapes of knowledge.[5]

In many cases, this probably doesn't matter. The fact that one may use a keyboard to type an email to the parent–teacher organization agreeing to bring brownies to play practice doesn't affect the message all that much. However, if one takes Kress's admonition to heart, and entertains the notion that screens and keyboards "will have profound effects" on learners' "engagement with the world," then it pays to explore what those effects might be.

The slightly peeved email capitulating to the onerous demands of the PTO for baked goods aside, writing as a form of communication and connection has played a crucial role in the development of just about every sphere of modern life. Anne Mangen and Jean Luc-Velay, researchers from Norway and France, respectively, provide this apt definition of writing: "A highly sophisticated and comprehensive way of externalizing our thoughts, giving shape to past memories as well as future plans and dreams, sharing our stories and communicating our emotions and affections."[6]

It's rare to read academic research that makes the heart sing, but something in their description does just that. Writing is important not because it catalogs how many goats some ancient Sumerian noble owned or describes disastrous foreign policy decisions in a 280-character tweet, but because it can be used to give shape to dreams, to open up the infinite interiority of our experience for others, to tell stories that give our days meaning, and to communicate our affections. Those seem like a necessary enterprises for a worthwhile life.

But Kress suggests that which technology is used to weave these stories will have a huge impact on "forms and shapes of knowledge."[7] When those forms are the stories told to one another about humans' experiences, and the means by which affection and understanding are expressed, then it matters very much what that the effect will be.

Many elementary school students in the United States now spend more time on laptops than writing by hand. This shift is prevalent not just within the American educational system, but globally. People write less and less by hand, something on full view when one walks into any Starbucks, where everyone is bent over a laptop writing the great American novel. The net effect of all this typing versus scribbling is unknown, but certainly there will be ramifications in how society as a whole views the written, printed, and screen-based word. But the other piece of Kress's observation is that this shift in technology will also have an effect on "bodily engagement with the world."[8]

One particular study highlights a growing observation made by researchers looking at the effects of the transition from writing by hand to writing on the screen. A study published in the journal *Psychological Science* found that students who took notes by hand versus on a keyboard understood concepts better.[9] Students who took notes by hand had fewer notes at the end of the lesson—hardly surprising as we can type faster than we can write by hand—but on later tests understood abstract concepts and ideas more clearly.

The study breaks the data down even further. Factual information retention was about equal between the keyboard and handwritten note takers, but retention was another story. Students who took notes by hand retained the information longer. These two central findings—students understood concepts better and improved memory through longhand note taking—are particularly telling.

What happens when you take notes longhand? It's alchemical, really, for the simple reason that you can't write fast enough to take notes verbatim. You have to truncate the ideas into snatches and blips of information. Listening to a lecture, the work you do is to synthesize thoughts and ideas, compress more rambling narratives into concise points and markers that can be jotted down. To take lived experience, the thoughts and gesticulations of the lecturer, the ideas and conceptual underpinnings of their narrative, and translate it into bullets of information is no easy thing.

What is fascinating about taking notes is the manner in which the note taker assumes a position of benevolent editorship; no, you say, I think what you *meant* to say was this. And you write down your version of events, you co-opt the narrative and create the story in your own words. In doing so, you take ownership of the information, bend narrative and facts to your will.

On the one hand, this is done for expediency's sake; you can't write fast enough to get everything. And on the other, it is done because writing allows you to take control of the environmental narrative (that which is going on around us) and refashion it in a way that suits your own peculiar perspectives. The authors of the study call this "synthesizing and summarizing content."[10]

There's also something to be said for the mystery of opening up your notes after a passage of time, and looking back at the record of your days. There, embodied in ink or graphite, is the scrawled perception of your former self. It is deeply interesting to look back at notes and see what state the note taker was in during a particular moment. Many students (and adult learners) struggle with remembering details and information—and so it's with a deep fascination as well as a utility that the learner can review the thoughts of a particular experience. Because of the "synthesizing" that you've done, what you get is a reworked version of events, a narrative you've created of your own history.

Digital notes cannot provide the same window into the past, into your experience. And this reckoning of self comes from the embodiment of learning in the form of physical note taking. How else to mark the passing of days, the time you've spent, if not by etching, by hand, in physical space, a marker that you were here?

Since the earliest days of human habitation, humans have been painting cave walls, scribbling madly to keep track of weather, tides, and the human heart. Proponents of the digital medium will say that the work continues, merely online—this is the sort of digital boosterism that claims more information is uploaded on the internet in a day than can even be conceived of. This is undoubtedly true, but the physical record is something altogether different. It is the "narrative self" made real, the transformation of experience into prose. Through this process of translating experience onto the page, you become something more than passive recorder of events. You shape narrative. You become a story teller.

NOTES

1. Zadie Smith, "Dance Lessons for Writers," *Guardian*, October 29, 2016. Retrieved from https://www.theguardian.com/books/2016/oct/29/zadie-smith-what-beyonce-taught-me.

2. Stephen King, *On Writing: A Memoir of the Craft* (New York: Scribner, 2000).

3. Elizabeth Gilbert, *Big Magic: Creative Living beyond Fear* (New York: Riverhead, 2015).

4. Frank Wilson, *The Hand: How Its Use Shapes the Brain, Language, and Human Culture* (New York: Pantheon, 1998).

5. Gunther Kress, *Literacy in the New Media Age* (New York: Routledge, 2010), 1.

6. Anne Mangen and Jean Luc-Velay, "Digitizing Literacy: Reflections on the Haptics of Writing." *Advances in Haptics* (Jan. 2010). doi:10.5772/8710.

7. Kress, *Literacy in the New Media Age*.

8. Kress, *Literacy in the New Media Age*.

9. Pam A. Mueller, and Daniel M. Oppenheimer. "The Pen Is Mightier Than the Keyboard." *Psychological Science*, vol. 25, no. 6 (2014), 1159–68. doi:10.1177/0956797614524581.

10. Mueller and Oppenheimer, "The Pen Is Mightier Than the Keyboard," 2.

NINE
Deconstructive Criticism

A large part of teaching, both in K–12 as well as in higher education, is centered on providing feedback to students. Particularly for writing instructors, this process can be incredibly helpful for assisting students in developing a clear and intelligible voice on the page. However, there can be times that students struggle with hearing feedback, particularly when it's designed to address the shortcomings of their work, or is structured as a suggestion of what to do better. "I just can't take negative criticism. I'd rather not hear it," they may say.

One of the goals of education should be aiding students in developing cogent and solid arguments, and an essential part of that is providing criticism. It seems, sometimes, that hearing criticism—particularly in light of the constant affirmation many students have become used to through social media—is harder and harder for them to take. But writing—and on a larger scale public rhetoric—doesn't work that way. *Life* doesn't work that way. Reasonable people can, will, and should disagree. But when we think about the physical manner in which negative stimuli is interpreted by the body, and a particular student's developmental stage, things get a bit more complex.

Bashing millennials and Generation Z for their supposed inability to take criticism has become something of a sport. Too sensitive, too inured to feedback through their echo-chamber digital bubbles, too coddled by helicopter parents—choose your invective-laden joust. Although some of these sentiments may have some degree of truth, they come with a crucial footnote: these are, in spirit at least, the exact same critiques leveled at every generation by the previous one.

By now, it should only take a little self-reflective sensibility to note that older generations accuse the younger of being weak and soft. It's cyclical, and happens every time some older person chafes at someone young complaining about some slight, real or imagined. When students start to snivel, one has to realize that, although one can posture as much as one likes, no writer or creator of original work enjoys being negatively critiqued.

However, for students negotiating education, it's vital that they develop the ability to learn from critique, and also accept that getting blasted because of a poorly written essay is a necessary wayside on the path to developing skills as a writer and thinker. Negative criticism is a rite of passage in a way. When students present work and it enters into the critical debate, it becomes part of the public dialogue. And, more importantly, in defending arguments they've made, students develop the second-tier skills vital for any analytic thinker: refining arguments, offering clarification, and contextualizing their work in a broader conversation.

But there is no doubt that even if a young writer is able to recognize that the very fact that his or her work is being taken seriously through critique validates their efforts, it's not unusual for these pleasant feelings of having *arrived* as a writer to be of short duration. Receiving feedback— for students, writers, teachers, anybody—can sting.

In fact, when getting negatively reviewed or criticized, it's not unusual to feel it as a physical sensation. Sucker punched, as if you've gotten the air knocked out of you, breathless; the list of how individuals describe feeling when receiving bad marks on a test or exam or paper could go on. An essay or research paper, like any large ambitious creative project, is the physical manifestation of hundreds—thousands—of hours of one's life.

Much of the time, teachers ask students to write work that is essayistic, weaving research with personal anecdotes and beliefs. This means the work students produce often contains a good deal of them. Like J. K. Rowling's Voldemort, students impart a piece of their soul into their work (if they take it seriously). It becomes a horcrux that contains a part of their persona. To attack their work is to wound them, they think. What students relate feeling, anecdotally, is physical pain—a genuine sense of having opened themselves up, made themselves vulnerable, and then being betrayed. Sucker punched.

Popular philosopher (there are two words you don't see together often) Alain de Botton famously blew up and ranted at Caleb Crain of the *New York Times*, who wrote a negative review of de Botton's book, *The Pleasures and Sorrows of Work*.[1] "I will hate you until I die," de Botton posted on the critic's blog. While mildly apologetic and a bit abashed afterward, de Botton at least partially held his ground, defending his right to get angry at reviewers who—especially when writing for an influential publication like the *Times*—can make or break a book's success.[2]

The words used to respond to criticism such as *injury*, *insult*, and *hurt* are, at their core, bodily metaphors. In fact, the etymology of *insult* can be traced back to the Latin word *saltire*, or "to leap." To be insulted is to be pounced upon, bodily, by reviewers or critics or teachers marking up an essay.

IMPENETRABLE

It's possible that the United States is experiencing a cultural shift in regard to criticism and feedback. In part, the blame may be with social media. In some cases, the internet is a cesspool of vitriol, trolls cyber bulling and negativity abounding, but for many young people the internet is also a constant stream of "likes" and confirmation. This bizarre duality can leave even the most confident person off balance, as the daily blast of social media feedback can turn on a dime and either provide empty salve for the spirit or cruel invectives and abstracted insult. But beyond the echo chambers of the internet, there are other ways in which the idea of criticism and critique, and the ability to receive it, may be changing.

The most visible representative of the American public, the president of the United States, just can't take criticism. When critiqued, Donald Trump responds with wrath, demeaning his critics or flat-out claiming that what they write is "fake news," even when it's backed up by facts, evidence, and empiricism. While demonizing the press is an age-old tradition among politicians, the vehemence and disdain that Trump exhibits for his critics is unlike anything seen in recent history.

To retweet an altered video of himself savagely beating up a man with a CNN logo superimposed over his head[3] is the grotesque posturing of a cartoonish bully, sure, but on a deeper level, it's also existentially chilling. All of Trump's damaging gaffs, categorically racist statements about foreign countries, "the Wall," anti-immigration policies, challenging the Mueller Report, the Stormy Daniels dustup, the Michael Cohen debacle, and vituperative tweets are of the same ilk—puffed-chest posturing, out-of-control backlash, and embarrassingly retaliatory. After all, being accused—even if only by suggestion—of conspiracy to interfere with an election and obstruction of justice is pretty serious criticism. If some basic level of disagreement is not honored as a part of democracy, then it's not a democracy. It's something else—autocracy, demagoguery, call it what you will.

Trump's refusal to take criticism is celebrated by his supporters. Trump's unwillingness to engage with his critics is likened to a kind of American bravado. By denying agency to what is viewed as the elite—journalists and the newspapers they write for, experts in the legal field, scientists, and military professionals—Trump is asserting the right of everyone to deny unpleasant, tangible evidence. It's permission to do you, as in *you do you*, a phrase author Colson Whitehead suggests is evidence of an "ever more complicated narcissism." It's a relief to do whatever one pleases without addressing the consequences. It's also, by medical definition, a sign of sociopathic tendencies.

The fear is that this impenetrability will trickle down to the rest of the citizens of the country. If this is the new normal, then constructive feedback is a thing of the past. It also makes one wonder: what is it about our bodies that either allows us to absorb criticism and use it constructively or causes us to freeze up in the face of negative feedback and makes us want to go all Hulk-like when we're criticized?

There seems little doubt that the digital generation—kids who've grown up with social media and a phone glued to their hands—suffer from an inability to hear negative things. But what's actually happening when the news hits anyone full in the face that something they've done, written, or said isn't all that great? What mechanisms get triggered? And how can an embodied approach to education take into account the way in which criticism—of an essay, a research paper, or a project—is received by students?

THE BODY NEUROTIC

Individuals who are more sensitive to criticism are usually those who rate high in neuroticism. One of the key symptoms of neuroticism is amped-up emotional reactivity. You're at the office, and someone notes that your desk is a bit of a mess, and you spend the morning fuming or anxiety-ridden over your inability to wash out your coffee cups on time and stack the papers you're grading with more aplomb. Not only do folks who may be neurotic have a stress response that is out of proportion to the stimuli, but they are more likely to suffer from depression and anxiety. They take criticism badly, and, just to make things as tough as possible, they also have the tendency to be highly self-critical.

What happens, exactly, in the brain and body when receiving negative criticism? This isn't small potatoes—school is nothing if not an endless cycle of feedback, assessment, and review. No one is safe; even teachers and professors get rated and evaluated by deans and staff. Even the internet has something to add in the form of review sites like Rate my Professor (according to which I am the "bitter . . . professor version of Jon Stewart," which is either insulting or a badge of honor, depending on how full my self-esteem tank is on any given day).

It's helpful to further define what criticism is as it relates to other environmental factors and stimuli. For example, a typical stress response recruits our body's fight-or-flight mechanisms, most prominently through our endocrine systems, notably our adrenal glands. Immediate, sudden stresses are dealt with pretty handily by various chemical responses in the body that increase heart rates, reduce cognitive load, divert blood to large leg muscles for running, and so on. Glucocorticoids—one of an array of hormones that respond to stressful stimuli—play a vital role in gearing the body up almost instantly to duke it out or haul butt outta there in the face of an approaching lion or angry footie hooligan.

Although criticism is also an environmental stressor, it does not necessarily employ the stress networks of the body, although it can. It's worthwhile to note that how the human body responds to negative environmental stimuli isn't clearly delineated between life-and-death matters and worrying about how your boss is going to yell at you for not finishing that report on time. So many factors are at play, including the environment we were raised in, how much access we had to adequate nutrition, genetics, culture, and so on, that to say that embodied responses to criticism and our body's response to direct environmental stressors (like lions, tigers, and Halloween mask–wearing psychopaths wielding knives) are different entirely would be inaccurate.

The human body has a finite number of systems to adapt and use to respond to our environment—criticism responses often mesh with stress in various ways dependent on the individual's history and genetic proclivities. In short, it is wicked complicated and what we're looking at is definitely just a piece of the puzzle, a narrow slice to examine to get a better grasp on the way feedback is embodied in education.

A number of reactions occur in the individual dealing with face-to-face criticism. The foundation for these reactions is the sensorimotor observation of the person doing the criticizing by the individual being criticized.

SMILE! YOU'RE ON *CANDIDE* CAMERA!

Candide, Voltaire's sardonic eighteenth-century novella, concerns the adventures of the eponymous hero as he journeys from a state of naive idealism bounded by limitless optimism to a begrudging confrontation with the nastier realities of the world. Like any eighteenth-century book, the going can be tough for modern readers, but one of the notable aspects of the novella is the way in which the desecration of the body and spirit go hand in hand. Bodies are treated horribly in the text. They are "thrown upon a dunghill," are "run through" by swords with startling frequency, and, perhaps worst of all, two naked girls are walking along "while two monkeys were pursuing them and biting their buttocks." These physical travails are emblematic of a deeper scarring; Candide slowly learns throughout the story that the world can be a pretty crappy place and his optimism may be misguided.

This marriage of physical and emotional pain is hardwired into the way in which the body interacts with the environment. When an individual sits and faces his or her interlocutors and receives bad news about a performance, a complex back-and-forth occurs between the critic and the criticized that begins with perception of the body.

The inherent pickle is that often criticism of the work is equated with criticism of the individual. This is what Trump suffers from; every attack is personal and revenge must be meted out accordingly. But this is exactly what one shouldn't do.

Buddhist philosophy and practice addresses this situation directly, and the lesson is often that a bit of mindful noticing of oneself in moments when being critiqued isn't such a bad thing. As author Bret Anthony Johnston stated, "Stories aren't about things. Stories are things."[4] That very thingness can help someone who is being criticized get the appropriate distance from their creation—providing the breathing room of healthy perspective. It's part of what teachers need to provide to their students if they are to become functional thinkers and human beings.

So what does one do with bad reviews? What are the embodied responses that do happen, and what are some alternatives? Sulking is mildly comforting, but not a long-term option. Getting apocalyptically angry, as de Botton did, is also available as a response, although de Botton had to apologize and retrench and ultimately try to rationalize his outburst, which is a whole lot of work with minimal return. Having tantrums, while refreshingly cathartic, is not particularly conducive to cultivating strategies to thrive as a public figure. Or as a writer, for that matter.

The reality of responding to bad reviews, or criticism more generally, in a way that offers at least a modicum of decency is not particularly sexy or dynamic. In the face of bad feedback, the writer, politician, or student has to quell subjective defensiveness and simply read the review on its own merit. Are there any truths lurking there?

Any reviewer worth his or her salt takes the job seriously. Very few are just out for a hit job, salivating eagerly over the takedown. It pays to read the reviewer's words, and take the bits that are valid and think about them. After all, even Dr. Seuss has gotten bad press. Reserve the right to disagree if you can muster a good enough reason. Maybe self-indulgently feel sorry for yourself temporarily; after all, you're not made of wood. Then, once ready, get back to work, and try to do better.

Given that there is an embodied reaction to negative feedback, there is also a way to divert the negativity away from an embodied response. Mindfulness exercises, whether body scans or relaxation exercises, have been shown conclusively to decrease stress. In addition, one of the results of sustained meditation practice is developing the ability to not take environmental scenarios personally. Yes, this is happening, you may say but not to me. It's just something happening. Buddhist meditation and philosophy brings about this point again and again, and as a means of absorbing criticism, it's vital. Mindfulness practice does, in fact, reduce stress levels, and if feedback is stressful, then regular mindfulness meditation may be able to mitigate the bodily response to negative criticism.

Or, perhaps, if you *really* want to take revenge, write a long chapter about criticism, turning the negative review into a positive gain.

NEGATING THE SELF

Conventional wisdom would seem to suggest that absorbing criticism in an educational environment requires only a healthy self-esteem and a tacit understanding of the milieu; yes, a student says, I'm being critiqued but that is a part of the this whole "learning" process. It is not about me, per se, but about my academic performance. For anyone who has been in a highly critical environment, however, maintaining such an objective distance from the criticism received is easier said than done. Educators are aware of how negative feedback can affect students and themselves emotionally. No one feels particularly jazzed about being told she's doing a crappy job at her job.

How does mindfulness meditation actually help with this scenario? First, it's useful to identify how it is that a practitioner of Buddhist mindfulness actually checks in on the "self," and who that particular "self" is.

Normally, you assess your state of mind, emotions, and feelings in a rather disembodied, psychological way. When answering the question, "How are you?" you catalog the things that have recently happened to you: "Pretty good, actually, my students seemed really into the lesson I taught today" or "Not great, my teacher was totally mad at us for not completing the project on time." What you need to realize is that these scenarios are representative of a narrative of events and causations that you have woven out of thin air; they are disembodied in the sense that they don't exist as empirical reality.

Interoceptive awareness, however, is a mitigating factor in this dynamic. Mindfulness meditation is the active embodiment of interoceptive awareness—or, put another way, if you *really* want to know how you *feel*, developing the ability to "check in" with your own corporeal self is a way to accomplish it. What you can discover is astounding.

The feeling you get during times of negative environmental stimuli can be counteracted by developing your interoceptive abilities. As much as it hurts to be told that your idea is foolish, the act of self-awareness that comes with mindfulness counteracts such narratives. But not in some cognitive, analytic way. Simply by engaging in a regular practice of mindfulness meditation you can stand a bit out of the way of yourselves and not let the bastards grind you down, to paraphrase Margaret Atwood.

Students would benefit from developing this ability. Contemporary educational environments are a barrage of negative stimuli. By reinforcing a sense of interoceptive awareness, students can remove the sting from negative feedback and more appropriately and accurately view their context to move forward in a way that makes sense. In fact, using critique and negative feedback as an advantage is at the core of many disciplines similar to mindfulness meditation.

One of the disciplines that is grounded in exactly this philosophical approach is aikido, the Japanese martial art developed by Morihei Ueshiba in the early twentieth century. What is particularly compelling about aikido is the philosophical foundation of the discipline. Developed both in harmony with other fighting styles as well as in contrast to them, one of the central tenets of aikido is not doing harm to your opponents—rather, a practitioner uses strength against the opponent and deals with both physical and emotional attack in as nonconfrontational a way as possible. Aikido sensei often tell little stories or offer up parables of how aikido works in life. An argument on a bus over available seating is a perfect place to use the philosophy of aikido, for example.

The word *aikido* is roughly translated as the "way of the spirit of harmony." Aikido involves many of the same techniques of mindful body awareness that we find in Zen Buddhism, and in any given studio, the sensei is always talking about balance, about calmness in the center of chaos.

One of the most noteworthy aspects of aikido teaching is the way in which students are taught about how anger makes you rise up. For example, much of aikido training is centered on throws and pins, techniques used to defend against attack. A central tenet of all of these techniques is that it's important to be solidly balanced, with a wide stance, to complete the series of motions the move requires. When teaching these techniques, the sensei will often observe how anger, fear, and resentment make the body rise up, how agitation and anger and resentment can place the body in a position where there is no balance, ready to topple over. This can be contrasted with bodily calm, where the individual is rooted in the ground, calm, alert, and ready, with a solid, wide stance.

Mind follows body. This is true in aikido, where the mind must be calm to deal with attack, but also in education, where incorporating negative feedback only works if the mind is able to receive the constructive elements of feedback. To appropriately deal with the attack, one must have a calm body and calm mind. If one of the goals is to prepare students for the "real world," then teachers must recognize all the places—school, work, in relationships—when emotionally charged moments can create a lack of harmony and take away balance, and how the steady, bodily work of regaining a center, becoming rooted, actually allows a person to move through the next moments of their lives with a sense of calm.

NOTES

1. Caleb Crain, "Toil and Trouble," a review of *The Pleasures and Sorrows of Work*, by Alain de Botton, *New York Times*, June 24, 2009, Sunday Book Review, https://www.nytimes.com/2009/06/28/books/review/Crain-t.html.

2. Caleb Crain's blog, with de Botton's post: http://www.steamthing.com/2009/06/review-of-alain-de-bottons-pleasures-and-sorrows-of-work.html; here is de Botton's Twitter message, where he first held ground: https://twitter.com/alaindebotton.

3. "Donald Trump Posts Video of Him 'Beating' CNN in Wrestling," *BBC News*, July 2, 2017, https://www.bbc.com/news/world-us-canada-40474118.

4. Bret Anthony Johnston, "Don't Write What You Know." *The Atlantic* (Feb. 19, 2014), https://www.theatlantic.com/magazine/archive/2011/08/dont-write-what-you-know/308576/.

TEN
Monkey See, Monkey Do

One of the rationales provided for education is that it provides students with "socialization." In some ways this is accurate, but in other ways it is misleading. Compulsory education in its standard form places large groups of children of the same age together for most of their schooling. Rarely in life, whether within communities, families, or the workplace, will such an unusually hegemonic structure be found. Most organizations and discursive groups that form and reform ad hoc through life are composed of individuals of different ages and backgrounds. Developmentally, it's not clear that simply because a student is thirteen he is ready for pre-algebra, or that a first-year college student is ready to write a comprehensive annotated bibliography at nineteen years old. And yet the modern educational system presupposes they are, despite the fact that human development is punctuated by fits and starts, stalls and sprints.

In addition, another drawback in placing students in same-age cohorts throughout their schooling is that it can reinforce some of the more negative developmental aspects of socialization. Particularly during adolescence, this can create a stressful environment.

Nowhere during the developmental stages of childhood and adulthood are humans more primed to "fit in" than during adolescence. This makes sense, from an evolutionary perspective. Modern primates form social groups similar to what anthropologists believe humans' hominid ancestors did. When individuals in these groups reach adolescence, the need to fit in is a matter of survival. From an embodied point of view, being in the crowd means protection, belonging, and the potential to find a mate. Being outside of the circle of peers makes a foraging primate susceptible to predators and less likely to create the necessary social bonds for full participation in group activities.

Teachers can see this frequently during both class discussions and classwork, where students unintentionally mimic one another. This can happen on multiple levels. For example, if one group of students decided to design a PowerPoint for their project, many other groups will follow suit. During a class discussion, if a socially high-ranking individual (popular kid) voices an opinion about the topic for the day, it's likely that others will quickly fall in line and agree. Particularly with adolescents, it's easy to see the way in which social cues inform a great deal of behavior. This tendency is exacerbated exponentially through social media, a kind of echo chamber of adolescent whims.

Acting out in class is usually a group activity, or designed for social effect, good or bad. Regardless, there is a constant interplay of behavior and rank, inclusiveness and exclusivity, operating in any traditional classroom. And despite the efforts of educational institutions to convince the public that what they're striving for is the creation of innovative thinkers and original creators, much of the time traditional school settings desire obedience from students.

Obedience from a behavioral perspective is based on physical mimicry. It's an embodied representation of learning during early human development. Infants of English-speaking parents breathe differently within hours of birth than children of French-speaking parents—just hearing the language surrounding them changes the embodiment of breathing. When humans are infants, much of their development comes from proprioceptive experiences. Many parents of newborn infants put soft mittens or socks on their baby's hands, as newborns have not yet developed the proprioceptive awareness to avoid scratching their own faces and eyes with their hands. This passes quickly, however, and within an incredibly short period infants, and especially toddlers, begin to clue into the world around them.

During this phase of development, infants and especially children begin imitating the people around them. This is, of course, a wise move from an evolutionary perspective. When a primate troop suddenly starts hooting in panic and everyone leaps into the trees, those that don't follow suit end up as a lion's lunch. It pays to mimic.

It's very difficult to break with behavioral patterns learned in this way—very difficult to avoid obedience. Looked at in a certain light, this obedience is intensely embodied. One of the starkest examples of this is the way in which institutional space is designed to limit movement. Schools, especially, are designed to move large groups of students to and fro throughout the day in as orderly a manner as possible. There is little in modern educational architecture that reinforces free movement, autonomy, and bodily exercise within the buildings; engagement with the body has been divorced from the structures of education and moved out to the athletic fields. And, although we wouldn't at first pass say that those in school, including teachers and staff, are embodying obedience, they are.

Individuals engaged in a traditional school environment are obeying a very strict set of social guidelines about how to move around and interact with the environment. The counterargument may be that the way individuals walk and sit and move is simply a result of morphology, seeking the most efficient way to orient the body to get work done. However, one gets a very different view of embodied learning upon visiting a halfway liberated third-grade classroom. Students there, if given the freedom, drape themselves over the backs of couches and read with their heads dangling off the side, sprawl on floors, walk over patterns in the rug while repeating their seven times table, roll around on carpets the way a dog rolls in the snow, curl up under desks to read *Junie B. Jones*, and generally resemble lolling baboons more than tidy office drones.

Obedience goes beyond physical mimicry, however. Multiple research experiments have strongly supported the idea that we are psychologically primed to obey authority. The most famous of these is the Milgram experiments by the eponymous Dr. Stanley Milgram at Yale in the 1960s. Participants in the study used electricity to shock individuals in another room with a painful jolt every time they answered a question wrong. The folks getting shocked weren't actually being electrocuted—they were pretending—but the person asking the questions and doing the zapping didn't know that. The shocks grew in intensity as more and more questions were asked, and the researcher running the experiment told the shockers that the final level of electricity could be fatal.

The majority of people—normal, decent humans—were willing to shock the hell out of folks even though they could hear screams and begging from the other room. Many were upset about it—crying, begging to stop—but more than half went through with it and zapped the bejesus out of a complete stranger simply because a researcher in a white lab coat told them to.

OBEY!

Social conformity begins in the body. It's no surprise that adolescents in middle school are painfully, acutely aware of their bodies—the way they look and smell, how big or little, fat or thin they are. Humans are primed to *fit in*, not stand out.

It doesn't pay to be a white buffalo—it just makes it easier for predators to spot you. Evolutionary adaptation has conferred upon each species a set of behaviors that maximizes survival—or at least makes reproductive success viable. To change these bodily behaviors is evolutionarily risky—it generally doesn't pay if you're a furtive, delicious prey animal to go gallivanting around advertising your drumsticks (the fact that some animals do, like greater sage grouse and peacocks, is another story about the risks and rewards of sexual display, but that's outside the subject of this book).

Put another way, when social animals like adolescent lions or monkeys or humans are seeking to find their ranking in a hierarchical group, it's wisest for them to follow the lead of those around them, particularly those ranked higher in the group.

The old black-and-white videos of goose-stepping Nazis are deeply disturbing and chilling. There they are, marching in perfect synchronicity, jackbooted and in perfect step. But it's emblematic of how scholars have tried to explain the horrors of the Holocaust. German Nazis were just following orders—being obedient. In the book *Hitler's Willing Executioners* by Daniel Goldhagen,[1] this concept is brilliantly articulated. Both our hardwired behavioral adaptation to embody the physical actions of those around us in early developmental stages and a culture predicated on hierarchies and rule following go a good deal of the way toward explaining why seemingly normal people participate in truly horrible activities.

This discussion becomes particularly pertinent when we begin talking about how much we want students to exhibit innovation, creativity, and "new" ideas. Stepping out of obedience is harder than one thinks.

The need we have to fit in is driven in large part by our evolutionary history. As social animals, we are constantly watching and assessing what our other troop members are up to and following suit or responding, and much of this perception is of bodily states.

SOCIAL INTELLIGENCE

Nowhere is status deemed more important than during adolescence. The idea of "fitting in" becomes paramount during this developmental phase, and kids spend an inordinate amount of energy figuring out the social rules of their given situation to find a social niche for themselves. How they do this depends, but, generally speaking, humans are excellent at mimicry, the bodily imitation or reproduction of other's movements and actions.

As we read others' body language and movements, a tremendous amount of cognitive processing occurs. To discern and analyze intent and character from embodied movements is difficult, but we've been primed to do so throughout our evolutionary history. When a dog "play bows" by sticking her butt up in the air and lowering her head and shoulders to the ground, tail wagging and face grinning, she is communicating a nonaggressive invitation to play to others, and this includes humans. So even in an interspecies relationship, bodily movements are forever communicating back and forth.

Although there is a subconscious understanding of this phenomenon—the bodily imitation of others in order to grease the wheels of social interaction—humans don't normally assess it on the same grounds as other forms of intelligence simply because we are such a verbal species; most of us are dependent on spoken information rather than what is being communicated bodily. However, whether we realize it or not, a tremendous amount of information is passed between humans during social interactions.

What does this corporoceptive ability mean in terms of how humans relate, and more specifically, how learning works? For example, many educators recognize the validity of talking about emotional intelligence; could "bodily intelligence" be added to the growing list of the ways in which educators measure the capabilities of their students?

Body language matters. Imagine a scenario wherein a less popular kid is hanging around the cool-kid crowd, watching their every movement, drawing in the language being offered by their bodies. By incorporating themselves into that movement regimen, by blending in, so to speak, with their bodily hierarchy, the kid on the "outs" is going to have a better chance of being accepted by the students with more social capital. This is so deeply hard-wired—the need to belong and be part of the group—that to deny its influence on education seems a bit foolhardy.

Research has demonstrated that there's a direct correlation between how well individuals are able to read and reproduce bodily movements. A 2018 study published in *PLOS One*, the journal of the Public Library of Science, strongly suggests that how well individuals can read and mimic the bodily actions of others is directly related to their social intelligence.[2] In fact, not only does the ability to copy bodily motions matter, but the study also notes participants engaging in "anticipated action," where they begin the behavior before the person they're watching even engages in it. Participants in the study who scored high on mimicry also scored high on social information processing. Not only were they adept at copying the actions of others, but they also were able to intuit and predict behaviors based on nonverbal cues (such as facial expressions, for example).

How well you fare socially has tremendous consequences throughout your life. During school, from elementary sandboxes on up to college lecture halls, although the ideas of personhood and general respect are given their due, not a lot of energy in education is dedicated toward the development of individual behaviors. Cognitive processes and skill sets are given attention, yes, but running students through an Emily Post–style etiquette school is the last thing most educators would want to see. However, there is one way in which teachers may be hampering students' abilities to maintain healthy social relationships and engage in social groups in such a way that maximizes their social capital. Education, in general, is not helping them develop their interoceptive abilities.

Creating a curriculum that prioritizes mindfulness is one important step toward addressing the way in which embodied responses play a role in child development. Here's why: it has been shown that mindfulness increases interoceptive abilities. The more proficient one becomes at mindfulness meditation, the more "in touch" one is with one's interoceptive states, which are directly tied to emotions. Thus, the mindfulness practitioner becomes better at *emotional regulation*, a key ingredient of successful social interaction.

In a study published in *Frontiers in Psychology* by authors Carole Hooven and Cynthia J. Price, the relationship between bodily awareness, emotional regulation, and how well people are able to respond to given situations is made clear:

> Emotion regulation involves a coherent relationship with the self, specifically effective communication between body, mind, and feelings. Effective emotion regulation involves the ability to accurately detect and evaluate cues related to physiological reactions to stressful events, accompanied by appropriate regulation strategies that temper and influence the emotional response.[3]

The ways in which students are able to successfully navigate social situations, respond to the bodily cues of others, read the nonverbal expressions of those around them, and mirror that action appropriately all depends on interoceptive awareness. Think of a situation wherein a friend is telling a story about something terrible that happened. You usually compose your face in an expression of anguish or pain similar to that of the speaker; this empathetic response makes the person who's telling the story feel listened to. If you don't respond in kind—anticipating the facial reaction of the storyteller and copying their expression—you may be seen as unfeeling and antisocial. This would hardly increase your social capital, make you any friends, or encourage others to confide in you.

In any given social situation, members of a group are barraged with signals: nonverbal cues from those in the group, body language, what individuals are saying, the tonality with which they speak, the complexity of what they're saying. Members of any social group are responding to all of it, all the time, whether aware of it or not. Folks who are adept at parsing meaning from bodily movement and responding accordingly are more successful socially; they rate higher in social intelligence processing. The ability to respond appropriately in an emotional capacity depends a great deal on our interoceptive awareness. Emotions begin in the body and are signaled by the body; therefore, appropriately regulating them given the context depends on the sensitivity of internal states.

Oddly, in most schools, the default mode is to sit students in rows, at desks or tables, cutting out the possibility of the kinesthetic conversation that could be occurring. The movements and physical cues that happen in normal human interaction are muted in the classroom, depriving children of the opportunity to hone the crucial skills of interpreting meaning and anticipating the actions and feelings of those around them and responding accordingly. The more educators can do to lift students out of the sedentary models of learning and engage in movements the better.

In fact, students who may excel socially (think of the social butterfly who can't concentrate on work but is well-liked and socially connected) aren't give the opportunity to use that skill to their advantage; instead, they are forced to perform a narrow set of tasks deemed appropriate for more traditional learning modes. But that charmer in the back row, who's got everyone laughing, that kid may have a combination of embodied intelligences that end up being far more beneficial in the long run than any particular lesson.

NOTES

1. Daniel Goldhagen, *Hitler's Willing Executioners: Ordinary Germans and the Holocaust* (New York: Knopf, 1996).

2. Oliver Genschow, et al. "Mimicking and Anticipating Others' Actions Is Linked to Social Information Processing." *Plos One*, vol. 13, no. 3 (2018). doi:10.1371/journal.pone.0193743.

3. Carole Hooven and Cynthia J. Price, "Interoceptive Awareness Skills for Emotion Regulation: Theory and Approach of Mindful Awareness in Body-Oriented Therapy (MABT)," *Frontiers in Psychology* vol. 9 (2018). doi:10.3389/fpsyg.2018.00798.

ELEVEN
Gut Feelings

Emotions and feelings don't exist in a vacuum. The manner in which mood and emotional states are expressed is contextually dependent on what is happening in the body, and perhaps the most important bodily connection is that between the brain and the gut. It's important to understand emotional states as not just expressions of internal physiological conditions, but as inextricably bound up in biological processes. The danger in categorizing emotions as "expressions" of physical states is that that particular conceptualization creates a causal relationship, whereas in fact the effect runs both ways. The way in which emotions manifest can be based largely in physical conditions in the body, but by the same token cognitive emotional states can drive internal, biological conditions.

The relationship between internal states and the outward expression of emotion is clear to anyone who has had to negotiate a particularly stressful environment when hungry. Here's the scenario: a student finally arrives a bit late to school, and due to a busy, chaotic morning, she hasn't had a chance to eat breakfast. School is demanding multiple things from her at once; she's on a deadline for a project due in Global Studies at noon but also has been pulled into a stressful social dustup over some thoughtless comments made on someone's Instagram post. As she tries to squeeze in a few minutes in the library to wrap up the project, one of her classmates leans across the table toward her to relate an annoying story involving comic romantic blunders, group texts, and upperclassmen that wanders and has little point. The overwhelmed, harried student feels frustrated and snaps rudely at her library tablemate, feeling bad about it but also unable to contain her response. That emotional response to a scenario that normally wouldn't have become so contentious is likely based on the lack of glucose in her bloodstream. Skipping breakfast has now led to a deepening of social anxiety, complicated interpersonal relationships, and a spike in cortisol levels that are already surging because of the Global Studies project.

During times of extreme stress and activity, the muscles and brain have to be supplied with glucose. The human body is great at figuring out how to burn fuel. When exercising, the glucose in the bloodstream goes straight to powering the cells in muscles. When not expending energy, glucose gets stored in fat. It's a super-efficient system that maximizes the return on calories.

Humans can actually bank energy, which is pretty amazing. But the problem is that human bodies evolved their metabolism during a large time span in which our evolutionary ancestors lived radically different lives. Although there are a few hunter gatherer bands still on the planet, most individuals in industrialized societies live within easy reach of some supermarket provender, which means that the energy output-to-input ratio is skewed. Hence, we have developed obesity, diabetes, and all the other health issues that relate to modern, industrial diets.

Feelings, actions, emotions, the ability to think, and behavior are all inextricably linked to food and the way the body processes it. Locating food through foraging or hunting in challenging environments has been a major concern for most of human evolutionary history, so it makes sense that many of the body's other systems—those that regulate and respond to hunger, satiation, emotions, energy levels, feelings—are tied to access to food. In fact, the phrase "eating my feelings," which people use as a joke to explain what is commonly known as "stress eating," is an apt description of the way in which emotions and food are intertwined.

You're feeling depressed, bummed out, frustrated, so you sit down and plow through a bag of chocolate-covered pretzels. Your spouse catches you, and you lamely joke, "Just over here, eating my feelings." And the truth of the matter is that your emotional state is tied to your hunger–satiation cycle, mainly through the vagus nerve, which is the neural pathway between your gut and brain, but also through the way your brain demands a huge amount of energy in the form of glucose.

When you're hungry—out of homeostasis—your interoceptive abilities are calibrated to send signals to your brain and behave accordingly.

Glucose in times of intense activity is a good thing. The state of the body, in terms of how many calories are whizzing through your system at any given time, has a direct influence on mood, memory, and the ability to learn.

Long-distance runners eat those little packets of sugary goo, and other endurance athletes often consume bread and pasta to "carbo load" before a race. Although diets and training regimens are as diverse as ever now — some athletes swearing by low-carb, high-protein diets, some by pre-game naps, and others by not washing their jockstrap all season — one thing they all agree on is that food is directly tied to performance. One of the most radically helpful educational infrastructure advances the American education system has made is to offer free breakfast for students in elementary school.

Part of the reason diet and learning are so tied to behavior is due to the balance, or imbalance, of the thirty-nine trillion microbes that live in the human body, most in the gut. A whole slew of recent studies has strongly supported the concept that the health of the human microbiome is instrumental in health and well-being.

BELLY BUSTIN' BACTERIAL BONANZA

Of all the emerging science about the human body that's plopped itself into the frame of reference lately, none are enjoying quite the level of notoriety as the gut microbiome. The gut microbiome is the digestive ecosystem in each person. Depending on region, diet, and other factors, each person on the planet is host to a complex ecosystem of bacteria and other microorganisms, primarily in the large intestine. These organisms—the microbiome—depend on their human hosts for food and their environment. In short, recent studies suggest the bacteria, viruses, and microbes that live on and in human beings have a significant role to play in overall health, behavior, and evolutionary adaptation.

Humans have approximately thirty trillion cells in our body, and about the same number, plus or minus ten trillion or so, of microbes. This means that paying close attention to the way in which bodies react and respond to the microbiome is key to understanding how human beings relate to their environments as well as their own interior compositions, and how those relationships bear on development, learning, and memory.

At its core, embodied cognition posits that much of behavior is dictated by, processed by, and originates within the body through systems that include the central nervous system and brain but also structures such as sensory organs, skeletal systems, and the digestive system. An embodied view of behavior must look at how the body relates to the trillions of demanding bacteria in and on human bodies. While many of them are symbionts—think of the *E. coli* in the gut (large intestine) that helps break down food—some are not. And all are pursuing their own ends: they are trying to reproduce and win the evolution game. To do so, some have developed intricate relationships with the human body and choreographed complex dances of human behavior.

Let's look at an example of the way these microbes play a role in behavior. Imagine a student named Jill. She's a high school sophomore, let's say. Smart, capable, and ambitious, Jill has the raw stuff necessary to be a good student. But she comes from a broken home. She's experienced her parents' divorce and lots of arguing; let's throw in some verbal abuse and financial insecurity as well.

Because of a decent amount of daily stress, Jill has grown up in a state of chronic stress response. That means her body is constantly primed for fight-or-flight behavior. Doors slam, voices are shouting, and there's a constant risk of major conflict in Jill's home life. At school, her AP classes are intense, and she doesn't have the financial support that socioeconomic stability provides. She doesn't have access to tutors at the country club to help her through the rough spots of her learning. So when teachers start barking at the class to emphasize the *importance of this test*, it sends Jill's system into overdrive.

Stress—as if anyone living in the twenty-four-hour news cycle of the twenty-first century needs to be told—is the human body's evolved response to environmental stimuli perceived as dangerous. For example, you're at the beach, playing in the waves, and all of a sudden a dorsal fin breaks the water twenty feet away. Shining and black, the fin is cutting through the surf right toward you. Your endocrine system fires up your adrenal glands, shooting adrenaline and cortisol into your bloodstream, which is excellent because you need to get the hell out of the water. Glucocorticoids get released. Pituitary and adrenal glands are pumping. Blood vessels constrict, your breathing starts to hyperventilate, and your heart rate skyrockets. You swim for shore, arms pinwheeling like a Mississippi paddle boat on meth.

You barely make it, Jaws ripping at your heels the whole way. You lie on the beach, exhausted, every nerve and muscle jittery and frayed as you swear off aquatic sports for life. You feel like an overstretched rubber band and a burnt-out lightbulb all at once.

Human bodies are excellently adapted for just this scenario. Sudden, acute stress in the face of immediate environmental stressors are no sweat. Well, they're *sweaty*, but humans evolved for millions of years as a puny, one-hundred-pound primate with no antlers or teeth or claws to speak of. The ancestors of modern humans have been dodging hungry lions and other beasties for a long time.

What the human body *hasn't* evolved to cope with is the long-term chronic stress of situations like a semester of AP English with a particularly sadistic teacher; grinding, multigenerational poverty; or the checkout line at Costco. Short, sudden bursts of acute stress are fine. Long-term, chronic stressors? Bad. Long-term chronic stress leads to heart problems and all kinds of health issues down the line. Human beings' Paleolithic forebears didn't have to stress about tax returns for months at a time.

So Jill's body is in a constant state of stress response. But what does Jill's stress have to do with gut flora, the microbiome, embodied cognition, and learning? The relationships among emotional regulation, health, and the gut biome are complicated, and there are still a tremendous number of unanswered questions. Stick with me, and we're going to wade into these complex waters together. I'll try to explain, you watch for dorsal fins.

Gut, Meet Brain. Brain, Meet Gut.

First, there's what is commonly known as the gut-brain connection. Scientists refer to this relationship as the *microbiota-gut-brain axis*. Although there is intense debate in scientific circles around how much the gut biome affects behavior and health, the connection seems clear. As a *Nature Review* puts it in an overview of the topic, "the mammalian gut microbiome affects the brain and behavior."[1] How much, why, or how is another question entirely. In the simplest terms, what's going on in the brain affects the gut *and vice versa*. The first part is something that is commonly understood: when you're headed to the job interview, your stomach is in knots; you get a bout of panicky diarrhea right before a first date; you have stomach cramps and jangling nerves as you get ready on your wedding day. But it also works in the other direction; when your stomach is upset, it affects your brain and cognitive abilities. The brain is great at interpreting bodily signals, but doesn't care if the stomach is upset because of a big presentation at work, because the digestive system is processing something toxic, or because there's a viral infection. It's all the same to the brain.

The vagus nerve is to blame. Next to the insula, it may be the best place to understand the body/mind connections within embodied cognition. The vagus nerve is a superhighway connector between the brain and gut. When the vagus nerve is severed in lab rats (someone call the rodent version of the American Civil Liberties Union) all kinds of crazy stuff happens in terms of behavior, immune responses, and health. Long story short: the vagus nerve is the main conduit of information between the brain and tummy.

Next, cytokines are a group of proteins in the body that signal cells. There are all kinds of different cytokines, but one particular group is part of the body's immune response: cytokines in that group cause inflammation. Inflammation is the body's all-purpose immune response. As an adaptation, it's pretty handy. It makes the area of infection or injury sensitive to touch by firing nerve endings (so the injured individual can favor it and let it heal — and know something's wrong), increases blood flow (so the blood can deliver all the healing chemicals and cells to the area of injury), and raises the temperature of the surrounding tissue, among other things. It's great as a short-term, general response to a physical, viral, or bacteriological attack — bad over the long term.

There are a number of other related factors in the relationship between the gut biome and behavior as well as health. The chemical responses that occur in the body as a result of stress and disease are (1) massively complex and (2) particular to a number of factors such as age, genetic disposition, sex, and on and on. But the factors at play in Jill's life are widely understood.

Jill has a lot on her plate and no support. So she's got a challenging environment at home, tough slogging at school, and all the BS social dynamics of the online world, and as a result she's in a constant state of stress response. Her body produces cortisol, which triggers the physical manifestations of stress.

But here's another thing that cortisol does: it makes her binge on junk food. She mainlines Pop-Tarts and shovels down sugary breakfast cereals. This intake radically increases the sugar levels in her body. So now, not only is increased cortisol pumping out cytokines that lead to inflammation, but also increased sugar levels spike insulin production; this releases cytokines, too. All that gluten she's eating—fine in normal amounts—gives her body a constant stress response, which is not great. This promotes a greater level of gut permeability. Also known as "leaky gut," this is a deterioration of the mucous lining of the epithelial gut wall that allows molecules to pass back and forth (the epithelial gut wall is a lining of specialized cells that help create a boundary between the body and gut microbes, and aids in the symbiotic relationship). Other factors can increase leaky gut as well, such as alcohol consumption. Gluten is also the favorite food of certain gut microbes in our microbiome.

Unfortunately, the bacteria that love chowing down on gluten, and multiplying when there's lots to be had, include pathogenic bacteria (nasty buggers that can make your colon sick), yeast, and fungus. (No one wants these in their bowels, especially because an increase in these particular gut flora trigger—wait for it—the immune-system response of even *more* inflammation.)

When Jill is stressed and eating poorly, she's causing havoc in her gut. Plus, she's got all that stress-related cortisol cruising through her bloodstream. At times of extreme stress, the human body does a great job mobilizing glucose for running away or fighting for survival. Muscles are jacked up with blood and adrenaline for escape or battle.

But during chronic stress, the body suppresses other functions to focus on only the fight-or-flight response. Cell division is costly, so that stops during stress, which explains why children who grow up in extreme poverty or in war-ridden regions often have health problems and low birth weights, and often don't grow as well as contemporaries from more industrialized, stable countries. Cortisol also depresses the immune system; after all, fighting off a cold isn't as important as fighting off a lion. Part of that now-stunted immune system includes the production of secretory immunoglobulin A, a vital front line of defense in the digestive system that regulates gut mucus. Jill now has a leakier gut than ever.

Jill's gut is now sick, sending a message via the vagus nerve to her brain that something's rotten in Denmark. (If Danes lived in her colon. Whatever, you get the point.) And here's where embodied cognition comes in. Her body is telling her she's sick. And because she's sick, her body has enacted the general immune response: inflammation.

Inflammation is recursive; it's self-perpetuating. Bodies respond to pathogens, injury, and infection with inflammation—when the body senses inflammation, it triggers an immune response that includes a second helping of inflammation. Those cytokine proteins cause damage to tissue and cells—even on the mitochondrial level—but they do something else as well. They switch the chemical messaging of tryptophan so that it no longer signals the production of serotonin and melatonin, neurotransmitters that, among other functions, regulate mood, sleep, and well-being. Instead, tryptophan signals the production of other compounds instead. The result? Exhaustion, messed-up sleep schedules, a drop in prosocial behavior, decrease in movement (exercise, yes, but also just general walking around, like how you don't go for a pleasant meander when you have the flu), and reduced libido. Finally, Jill's predicament has so impaired the functions of the gut-brain axis that *the very regions and processes central to learning and memory are impaired.*

Because of the way the brain relates to the gut via the vagus nerve, if things aren't balanced in the colon—if the body has fallen out of homeostasis in the gastrointestinal tract—it directly affects the brain's ability to remember, to learn, and to process new information. This works in a complicated way, but in essence the way it can affect behavior and cognitive abilities is as follows.

A diet high in sugary fats often leads to obesity. This is hardly news. But what is not as well known is that the consumption of such fatty, sugary foods also creates an imbalance in the gut microbiome, as certain bacteria in the gut favor such food sources and are able to out compete more benign, or even helpful, species of bacteria. Thus, the population of the gut microbiome becomes lopsided, something referred to as "gut dysbiosis" by researchers. The long-term effects of such gut biota imbalances include cognitive decline.

Researchers in Spain have explored the relationships among gut dysbiosis, obesity, cognitive decline, and mental disorders, such as anxiety and depression. In their article in *Frontiers in Neuroscience*, the researchers observed the following consequences of obesity:

> Common co-morbidities not only include cardiometabolic disorders but also mood and cognitive disorders. Obese subjects often show deficits in memory, learning and executive functions compared to normal weight subjects.... [O]besity is associated with a higher risk of developing depression and anxiety.[2]

The research into the gut biome is demonstrating that the health of those trillions of bacteria in the gut plays a vital role in mediating the relationships among diet, environment, behavior, and health in ways that can affect cognition and learning. One of the effects of long-term, chronic obesity is that it disrupts homeostasis, wherein the body maintains appropriate levels of glucose and regulates heart rate and temperature, as well as other factors. The system that handles the interplay of environment and chemistry necessary for homeostasis involves both the endocrine system (hormones like glucocorticoids and estrogen, for example) and neurotransmitters such as dopamine. Inflammation—a common side effect of obesity—can lead to numerous health issues, as the body struggles to handle gut dysbiosis and regulate itself. The problem of obesity and the disruption to a healthy gut biome is cyclical.

School often exacerbates the situation, through stress and a culture of junk food consumption, lack of sleep, and digital absorption—all of which have been connected to various dietary problems.

This is different from conversations around what an appropriate diet should be. Looking at diet alone and its relationship to learning is important, but other factors—namely stress and trauma and their long-term effects—must be taken into account as well. But if educational institutions are seeking to instruct the whole child and provide the sort of grounding for lifelong learning that many schools claim, part of that learning very possibly should include an understanding of the way diet and lifestyle and the gut biome can directly affect the ability to learn calculus, memorize the state capitals, and function in school.

GOING WITH YOUR GUT

One of the great running gags in modern cinema is the repetition of the line "I've got a bad feeling about this" in the Star Wars franchise. Han Solo says it, Princess Leia, Luke Skywalker, C-3PO, Obi-Wan Kenobi, Young Anakin—the list goes on. But what, exactly, do people mean when they're talking about a "bad feeling" in a given situation? What does one's gut feeling have to do with learning, and more importantly, with one's *actual* guts? Through the lens of embodiment theory, this metaphoric approach to describing feelings that may be outside the reach of articulation is closely aligned to the sense of interoception—uniquely tied to the gut—as well as intuition.

When the phrase "gut feeling" is used, what is often being talked about is that quirky scenario where an individual knows something but doesn't know how they know. Intuition, in other words. Gut feelings—about people, situations, outcomes, or environments—interconnect with various physical structures in the body.

For example, let's say your dean or principal calls you into her office. She sits you down, and states that you're being given a promotion. Yay! You think. You'll now be the departmental head or something. You're understandably ecstatic; this at first seems like a good thing. However, she goes on to describe the position, and you realize that you have to continue doing all the work of your old job, and the added responsibilities of the new position as well. You get a new office, but there's no pay increase. Great, isn't it? Asks the principal or dean, smiling.

No. Not great, you think. This sucks. It's *unfair*.

Humans' sense of what's fair or not is highly attuned. The very idea of "keeping up with the Joneses'" underscores this fact. Most people respond negatively to situations in which they are treated unfairly. In addition, they are able to perceive when others are being shafted as well, and there are some embodied reactions to these types of scenarios. Sometimes, when you say "I feel your pain, man," you really do in fact experience—on a neurochemical level—someone else's pain. This idea of fairness is what psychologists call "social equity."

The question is as follows: if a person is less embodied, for example out of touch with his or her interoceptive ability, is the sense of social equity compromised? To put it more succinctly, does an individual's sense of social justice depend, in part, on his or her own sense of embodiment?

Antonio Damasio's research is compelling in suggesting that decisions are bound up in unconscious biases, which precede any rational decision making. Damasio puts it this way in a 1997 article in *Science*: "nonconscious biases guide behavior before conscious knowledge does. Without the help of such biases, overt knowledge may be insufficient to ensure advantageous behavior."[3] In short, often people make decisions in large part based on their sensorimotor perception of past experiences. When faced with a decision to make, or a scenario in which fairness and equity are involved, humans have a finely attuned sense of what sorts of situations have been similar in the past, and the sensorimotor information encoded in memories of such events informs the choice.

Acting in a manner advantageous to the individual is hardwired into human's behavioral systems. However, it has been found that Buddhist meditators have the ability to override what might be considered "selfish" motives during research experiments. This makes sense; experienced mindfulness meditation brings with it a highly attuned sense of interoception. This suggests something that is at once obvious but also rather striking: humans don't all regulate emotions the same way.

Children in particular are obsessed with fairness, and don't have the regulatory ability to always keep their emotions in check. Just hand out birthday cake at a party, and watch the chaos ensue when you randomly choose one kid to get a massive piece with all the frosting. In that particular scenario, or any in which an individual feels slighted, the ability to resist the memory of that unfairness and recognize that in certain situations there are exceptions to the "fairness" rule should, theoretically, increase. But to be more in touch with a sense of social equity, one needs to be in touch with the interoceptive awareness of the body's response to situations that may bring up memories related to issues of fairness.

One of the goals of education is to create citizens—individuals who understand their reciprocal role in society. Developing a sense of fairness, of social equity, may be at least in part a matter of attuning students to their own interoceptive awareness and making sure that the decisions that they make for themselves and others acknowledge and take into account the complicated interplay between somatic memory, emotion, and action.

And oddly, there seems to be growing evidence that the sense of fairness that humans possess could have something to do with the trillions of microbes we play host to. Compelling research out of Tel Aviv suggests, through computer models, that hosts who have microbes that promote altruistic behavior out-survive those that don't.[4] This flies in the face of basic Darwinian natural selection theory, which suggests that all individuals act in a way that increases their likelihood of producing offspring—in a word, selfishly. But the science of natural selection is anything but straightforward, because there's also "inclusive fitness," wherein members of a species put aside their own selfish reproductive goals in favor of helping out the group. Think of worker bees in a hive, who don't reproduce. Altruism is working for the good of all instead of oneself. For example, a really good babysitter doesn't necessarily confer any reproductive benefit on himself for taking good care of a couple of kids. Altruistically, he does provide child care, however. Or maybe it's just to raid the fridge and make a few bucks.

Backing up for a second, many microbes change host behavior in a way that promotes their own survival at the cost of their host's life. Think of dogs with the rabies virus. The dog gets rabies, goes wacky, and starts frothing and biting other animals, thus spreading the virus. There's a fungus that infects the brains of ants and makes them climb up stalks of grass, where they zombify, cling onto the stalk, and eventually rain spores down on other unsuspecting ants. The natural world is full of parasitic microorganisms that change host behavior to the detriment of the host and benefit of the microbe.

At this point it's necessary to state that the field of the gut-brain microbe axis is heavily contentious. Many of the theories suggesting the role the gut plays on behavior have critics, rightfully so. But it's helpful to think about one example of how the gut microbiome could get Jill back on track. Remember Jill? A researcher at the University of California, Los Angeles, recently reported on how bacteria in a healthy colon encourage intestinal cells to produce serotonin.[5] Serotonin is vital for a healthy intestinal wall, but it's also the chemical that has been shown to alleviate symptoms of depression and anxiety. Prozac, for example, regulates how serotonin is processed in the body.

Although much of the research around the gut-brain axis is up for debate, what seems clear is that there is a connection between the health of the human microbiome and behavior and cognition. Exactly what that connection is, how important, and in what ways are questions that fuels late-night debates at nerdy science conferences around the world. But eating a healthy diet with lots of fiber, getting exercise, drinking a glass of water now and then, and avoiding giant bags of sugary snacks and sodas are things that most of those researchers would probably agree with.

NOTES

1. Katerina V. Johnson, and Kevin R. Foster. "Why Does the Microbiome Affect Behaviour?" *Nature Reviews Microbiology*, vol. 16, no. 10 (2018), 647–55. doi:10.1038/s41579-018-0014-3.

2. Ana Agusti, et al. "Interplay Between the Gut-Brain Axis, Obesity and Cognitive Function." *Frontiers* (Feb. 26, 2018). https://www.frontiersin.org/articles/10.3389/fnins.2018.00155/full.

3. Antoine Bechara, et al. "Deciding Advantageously before Knowing the Advantageous Strategy." *Science* (Feb. 28, 1997). https://science.sciencemag.org/content/275/5304/1293.abstract.

4. Elizabeth Svoboda. "Can Microbes Encourage Altruism?" *Quanta Magazine* (June 29, 2017). https://www.quantamagazine.org/can-microbes-encourage-altruism-20170629/

5. Thomas C. Fung, et al. "Intestinal Serotonin and Fluoxetine Exposure Modulate Bacterial Colonization in the Gut." *Nature Microbiology* (Feb. 2019). doi:10.1038/s41564-019-0540-4.

TWELVE

Virtual Race

Human beings use visual perception as their predominant means of interpreting experience. Dependence on visual information increases with age. While other sensorimotor information is also utilized to categorize experience—sound, feel, taste, smell, interoception, proprioception, and balance—vision remains the primary vehicle for environmental stimuli.

As such, it is through the subjectivity of perception that experience is defined. Vision is often mistaken for objective reality, but it is not. And the subjective nature of perception is nowhere more cruelly highlighted than in racist ideologies.

Slavery and the racially motivated genocide of the Native Americans were based on the perception of difference. That difference came in many forms—cultural, linguistic, social—but the most consistent feature of racism is the manner in which racist ideologies use subjective interpretations of physical appearance as the bedrock for hateful ideas. Racism also exists as an inherent part of American society—its philosophical underpinnings and the way it was institutionalized—and can be identified in the way founding documents of our country were written.

There are other ways in which racism is hardwired into American culture. Jim Crow laws and segregation were the most visible form of racism in the twentieth century, and other insidious structural prejudices exist in the way that, for example, urban planning isolated black communities, barring pedestrian access to downtown shopping districts. Lack of access to education and health care has disenfranchised people of color.

Of the more egregious examples of structural racism, one thinks of the recent news stories of black women whose doctors disbelieve their descriptions of symptomatic pain, a scenario that suggests that black women don't get access to the same type of health care as white women.[1]

Teachers and educators are on the front lines of these battles to confront racism as it exists structurally in institutions, including schools. However, often when there are diversity trainings at institutions or work on racism in education, there can be conflict. White participants don't want to be shamed, and to admit to bias is too embarrassing to consider. Many times, statements such as "I don't see color" mask the existence of a deeper bias that plays out in unconscious microaggressions. There's a helpful way to look at the manner in which certain individuals who belong to dominant groups (white, middle-upper class) in order to suss out why exactly they feel so attacked when their views are criticized. How hard it is for people who claim that they "don't see race" to understand that their experience isn't representative of anything other than their own, subjective, sensorimotor perception of the world?

Our perception of the world is embodied. This is why people who love chocolate or sushi find it hard to believe when others don't. Each person is caught in their own perceptual reality. Subjective interpretation of the world defines, in large part, feelings, emotions, and behaviors. Thus, in terms of race particularly, individuals experience their reality through the prism of their race.

For example, a white male experiences his subjective reality differently than an Asian woman, and that experience is dependent on context, different for one in Tokyo and the other in Akron. Following the line of argument offered by embodied cognition studies, this means that one's entire concept of self—not just one's own sense of being but the sense of others as well—is filtered through the subjective experience of race.[2]

In theory, humans can imagine what it's like to be a sparrow flitting on the wind, but for most people it's hard to really *know* what it's like to fly. Humans can't imagine sparrowness, much as they may like to try. Subjective perception of the world is such a powerful filter through which individuals internalize and process experiences, it's very hard for a white person to understand what it's like to be a person of color, whose very embodiment also filters their subjective experience.

This is unfortunate, but it gets worse. Turns out that if you're white, you filter the experiences of others and mete out empathy according to who belongs to the same race as you. A number of studies have shown that individuals have greater sensitivity to the observed pain of others, but only if those people are of the same observable racial category.[3] This strongly suggests that humans consciously—and subconsciously—identify most closely with others who have the same racial appearance. To put it bluntly: if you look like me, I'd be more likely to have empathy toward you.

This finding has been reproduced in multiple studies, and the research presents tough obstacles to diversity, equity, and inclusion trainings and work at educational institutions. Taken at face value, it confirms a poisonous notion: that humans are hardwired to see other races not similar to their own as "other" and that there is some embodied mechanism that separates humans into affinity groups based on race and apportions out the better parts of human nature in the form of empathy and compassion based on something as superficial as physical likeness rather than a shared humanity.

This inherent subjective perception can challenge an educator interested in moving the ball down the field in dialogues that address diversity, equity, and inclusion and intent on creating helpful experiences for students to combat this kind of bias.

IMPLICIT RACIAL BIAS AT THE DEPARTMENT STORE

Let's define an important term before we get cooking with the science here. *Implicit racial bias* is a banal-sounding string of words that belies its toxic meaning. Essentially, the theory is that humans subconsciously attribute traits to various races, usually in the pejorative sense (it should be noted here that this theory has come under fire because it is difficult to measure such attribution of traits).

Here's a brief example to back this up. Let's say there's a student of southeast Asian descent, India to be specific. She works at a local department store in a predominantly white state like Vermont or Wyoming. One day, a kind-seeming older gentleman approaches her while she works and starts chatting amicably with her at checkout. "Where are you from?" he asks. "Smallville," she replies, which is a little farm community just up the road nestled against the foothills. "No," he says. "I mean originally." There's a bit of a pause here, and she deadpans, "Smallville."

This is implicit racial bias: the assumption that anyone with brown skin must be from somewhere else if you live in a white state. A group of researchers from the Universitat de Barcelona in Spain devised a straightforward experiment based on our embodied sense of self to explore the notion of implicit bias. Again, humans view the world through their own visual perception of their race. When white people looks at their environment, they are seeing it through a lens of whiteness; that's their subjective reality. But what would happen if they looked down and saw brown skin instead?

The researchers ran participants in the study through a series of tests to assess racial bias. Turns out, most individuals had racial bias toward other races, as expected. Then, the researchers outfitted individuals in the study with virtual reality goggles, and adjusted the equipment so that when participants viewed their own bodies, instead of seeing white bodies they saw dark skin. Guess what happened? According to the authors of the study, "[W]e demonstrated that embodiment of light-skinned participants in a dark-skinned VB [virtual body] significantly reduced implicit racial bias against dark-skinned people."[4]

ENTER THE MINDFULNESS DRAGON

The brain has depended on the senses for the duration of human evolution. If humans' early hominid ancestors smelled the reek of hungry lions during their early days as scurrying east African hominids, it paid to take the information at face value and run. If an individual's hands pick up a plant that turns out to be a prickly cactus covered in spines, it is evolutionarily advantageous to not stuff it into their mouths without first dethorning it. When prairie dogs chirp an alarm, other prairie dogs in the town listen to the warning and scurry for shelter. It they don't, they may take a quick flight in the talons of a hawk or end up in a coyote's gullet.

Disbelief and skepticism are healthy in intellectual circumstances, but can be dangerous in the wild. Doubting your senses, in general, isn't a good idea. Eat something that tastes bitter and gross and nasty, and there's a good chance it's not going to end well for you. Ignore your senses at your own risk.

When your eyes—ensconced in virtual reality goggles—see your body as dark-skinned (if you're white), your brain is surprisingly flexible. Okay, your brain says, you're dark-skinned now. And then, all of your subjective experience of the world—the cataloging of your environment and the assemblage of meaning that follows—is filtered through this sense of self that now includes being dark-skinned.

There's something great but also a little sad about this research, as it suggests the path to correcting implicit bias may be accessed by something as (relatively) straightforward as virtual reality scenarios, but also that humans are hardwired to only care about their own "in" crowd, whatever that crowd may be. The experiment doesn't erase implicit bias; in a way, it *uses* it to trick the brain into being less discriminating, in this case, toward historically oppressed people with dark skin.

The effects of reduction in implicit racial bias after this type of experiment have been shown to last up to a week, suggesting that repeated exposures to this type of scenario could be effective in reducing racist attitudes—but there's nothing definitive in that direction yet in terms of research or conclusions. But it's worth drawing some connections to other ideas of embodiment to nudge the conversation forward.

For example, if experienced Buddhist meditators—who have increased interoceptive sensitivity—engage more deliberately in situations in which social equity is at stake, they have been shown to make choices that, although disadvantageous to themselves, are advantageous to the group. Is it not possible that a more highly developed interoceptive sense could also be used to mitigate implicit racial bias?

During a commonly used psychological study tool referred to as "The Ultimatum Game" (which sounds like a young adult novel), two participants are given sums of money. In one scenario, they are offered equal amounts and can both keep the money. In another, however, one is given more than the other, but the twist is that if one of them refuses, neither get the cash. Often, individuals will forgo getting any money whatsoever, as they see the inequity of it as too unfair. Why should she get more than me? Despite the old adage "some is better than none" many participants in the game will refuse deals in which they get less. This response has been shown to have a strong embodied component tied to the decision-making process.

Enter the dragon, if the dragon is an experienced mindfulness meditator, who tweaks the stats: a Buddhist proficient in the interoceptive art of meditation is more likely to accept a deal unfair to themselves. This is in part because of a deeply embodied response to inequality—individuals respond viscerally through a number of physiological systems to situations perceived as unfair—and mindfulness meditation engages individuals' very systems of interoception, and in doing so introduces a measure of control. Just because the body tells someone to do something doesn't mean the person has to listen.

It stands to reason that if educators worked more intentionally on developing students' interoceptive sensibility, it would make it more likely that implicit bias, which is an embodied reaction similar to the affront felt in unequal situations during The Ultimatum Game, could be reduced? It's hard to say given the early stages of the research. But given the way in which implicit bias appears to be a function, at least in part, of the embodied nature of the subjective perception of "self," exploration into the role of further developing interoceptive abilities as a means of reducing implicit bias isn't too far off the mark as a means for educators to begin designing effective trainings that address diversity.

EMBODYING DIVERSITY

Diversity work can be particularly challenging at majority-white institutions, where the burden of developing programs to address race issues often falls on the minority population of people of color. However, in higher education, at the very least, the idea of diversity, equity, and inclusion isn't anathema, and most institutions thankfully recognize it as a priority. Despite that, however, at the many institutions, faculty, staff, and student body struggle mightily with issues related to race and discrimination, diversity and inclusion. Just because people have PhDs doesn't mean they can't be prejudiced.

Many colleges engage in various forms of diversity-oriented trainings—some given by experts and consultants, some delivered by in-house folks. But, again, despite all of that hard work and money spent, there are, with regularity, racist incidents on campuses around the country. In the past few years, white supremacy posters have popped up around campuses in the United States, many following the white nationalist rally in Charlottesville, Virginia, in 2017.

The problem is that diversity training isn't the most effective tool to combat racism. A study by three researchers—Alexandra Kalev of Berkeley, Erin Kelley of the University of Minnesota, and Frank Dubbin of Harvard—looked at diversity training in hundreds of companies over decades.[5] The study showed that diversity training just didn't work in terms of removing stereotypes and advancing minorities in the workplace. "Diversity training sessions," "diversity performance evaluations," and similar programs didn't seem to have any positive results in promoting diversity and inclusion in the workplace. One of the aspects of this problem is white men. According to the study, "Laboratory experiments and field studies show that it is difficult to train away stereotypes, and that white men often respond negatively to training."[6] These types of trainings, say the authors of the study, have simply "not worked."[7]

So what does? The study points to two clear ways to improve institutional climates in terms of diversity: hire more diverse individuals for managerial-level positions and create mentoring programs open to minorities and women.

The first is a no-brainer. Hiring minorities and women for positions that have a certain amount of power has an immediate effect in terms of confronting and addressing diversity and inclusion in the workplace. For institutions that care about such issues, it is a case of "putting your money where your mouth is" and creating real, demonstrable change.

The second idea, mentoring programs, has also been shown to generate positive results. This, too, upon a bit of reflection, isn't surprising. It's hard to eradicate stereotypes with finger-wagging diversity trainings that may actually reinforce concepts of division, but active interpersonal engagement with other people often allows us to see people as individuals rather than categories. It breaks down the entire concept of "the other" that stands in the way of diversity work.

One of the other issues with formal diversity training is that the manner in which it is designed stands in direct contrast to what we know about embodiment. What we learn is intimately tied to *where* and under *what conditions* we do the learning. The context of our learning directly affects our ability to later recall what we've learned.

Context-dependent memory is an embodied form of cognitive retrieval. It's a pretty straightforward idea. The environmental context in which information is learned is tied to that learning, so when information is recalled at some later time, that recall is easier for the individual when he or she is located within the original context of the learning. For example, way back in the 1970s, when bell-bottoms and disco reigned supreme, two researchers, D. R. Godden and A. D. Baddeley, taught a list of words to a group of people in two different environments: on land and underwater. While underwater, the participants in the study wore scuba gear.

The study showed that "lists learnt underwater were best recalled underwater and vice versa."[8] (It also demonstrated that research done in the 1970s was groovy and weird.)

It should be noted that there is a whole library of studies on this phenomenon that both support and counter the basic premise of context-dependent memory. However, it seems likely that context-dependent memory is a verifiable trait in some cases, and that, as the authors state in their scuba study, "the phenomenon of context dependent memory not only exists, but is robust enough to affect normal behavior and performance away from the laboratory."[9]

Marcel Proust would've agreed. In his famous opening for the novel *In Search of Lost Time*, it is the bite of a madeleine cookie that sends him wheeling back in time to recall his youth—a sensorimotor experience (in this case the look, taste, smell, and feel of the spongy little shell-shaped cookie) that allows him to recall the events of the past.

From Proust and the scuba divers, a number of lessons can be brought straight to the workplace. If discriminating behavior is learned over the course of human development, in the messy moments of life as individuals progress from childhood to adolescence to adulthood—if this is the environment wherein stereotypes are learned and prejudices engrained—and then an attempt is made to train individuals in bland institutional settings like conference rooms and seminars where they just sit and listen, this scenario ignores the way in which embodied relationships to the environment embeds learning. While participating in a diversity training can help, it's unlikely that those who went through such trainings will be able to engage with that learning in a meaningful way back in the classroom or office.

There is no magic bullet to address racism in education. It's a hugely complex issue, from historically codified structural racist ideologies down to personal beliefs embedded in the subjective prejudice of the individuals. But one way to think about it is from an embodied perspective. Three ideas can help us move forward:

1. The subjective experience of race limits empathy for other races—but if a white-skinned individual can *perceive themselves* as dark-skinned, that limitation is mitigated.
2. The ability to recall information learned is tied to the environment in which the learning occurred.

3. Cognition, recalling information, and perception of race are all embodied, meaning those processes are collected, filtered, and assessed through interconnected networks of sensorimotor systems.

For this next part, imagine a scenario. You are putting in a new bathroom in your home. New tile floors, sink, toilet, shower. To understand how to replace major plumbing, you watch YouTube videos that show you how it's done. There are countless how-to videos on YouTube, offering tutorials on everything from installing toilets to tiling walls to replacing plumbing. However, even though you watch the videos diligently, when you actually set about replacing bathroom appliances and trying to tile, you find that the toilet leaks, the plumbing is backward, and the tile begins to crumble and crack. What happened?

You've fallen prey to a scenario that researchers from the University of Chicago documented in a study.[10] Turns out, watching YouTube videos that teach home repair, or building, or even throwing darts, has zero effect on your ability to perform the task after. What watching the videos does, authors of the study Ed O'Brian and Michael Kardas claim, is boost your confidence. In watching the YouTube videos about toilet installation, you've convinced yourself you know what you were doing.

But there's so much intuition and experience that come into play. A more seasoned plumber would've corrected the errors that you didn't even see. In the study, individuals who watched darts being thrown and hitting a bull's-eye believed they'd be great at darts, but turned out to be no better than participants who didn't watch the videos. But here's where things get interesting. Offer participants an *embodied way to engage with the task at hand*, and their sense of false confidence fades.

For example, if folks in the study watched a how-to video of someone juggling bowling pins, they thought they'd be able to have some degree of success. But, when they were able to actually hold a pin for a bit after watching prior to attempting to juggle, their expectations of success were more realistic. Our bodies understand our limitations better than our brains. The embodied, tactile nature of the world—in this case a weirdly shaped bowling pin—reminded them of the often difficult-to-manage physics of spinning multiple objects in the air and not dropping them. Long before our brain kicks into gear, our body has already perceived, assessed, and calculated the odds for success.

In an admittedly roundabout way, this brings us back to diversity training. If education seeks to address racial inequality in the classroom and in the faculty break room, attending diversity trainings alone won't cut it. Just like watching YouTube how-to videos, these trainings don't take into account the embodied reality of learning, or context-dependent memory.

It seems likely that hands-on, in-situ experiences around diversity and inclusion would be beneficial. Just as wearing virtual-reality goggles that darkened the appearance of skin increase empathy for others with dark skin, diversity work that asked white participants to embody the experience of people of color has the potential to be more effective. Racial discrimination is an embodied reaction to individuals we categorize as "different," not an abstract, reasoned and rational belief. Some of the nicest, most well-meaning, well-educated white people commit microaggressions all the time. These are the same people who've attended multiple diversity trainings and even work in professional capacities in positions aligned with diversity, equity, and inclusion efforts. Any white teacher who has taught a diverse classroom has undoubtedly made a vast number of unintended gaffes related to race, ethnicity, religion, and gender.

But it's in the moments of actual experience—when students of color feel hurt and antagonized by racist antiimmigration policies, female students are crushed by the nomination of an accused sex offender to the Supreme Court without due process, or trans students watch as their right to join the military erodes before their eyes—that's when individuals who are white and male have to realize that their own subjective experience is exactly that: their individualized, perceived reality of the world. *None* of these issues are going to affect how a white, heteronormative, middle-class American male gets through the day. Not a single one.

What educators who belong to the white majority need to do, then, is not try to equate their own experience with that of disenfranchised minorities (one of the ways in which white people attempt to sympathize is by comparing the experience of their white immigrant ancestors, believing that if their grandparents were poor Ukranian Mennonites, they understand oppression. However, this is an incorrect, apples-and-oranges analogy.) Rather, white individuals have to try to empathize with others' experiences through actively embodying their reality in some way. Clearly, that is easier said than done.

Formal diversity trainings are a good idea. They represent a structured acknowledgement of issues facing organizations. They also cost money, which is important: schools and colleges need to invest in work that directly faces up to discrimination in the classroom and workplace. The research suggests that that's not enough, however. For white people to gain traction on the slippery slope of their own bias, opportunities must be created for them to engage bodily with the lived reality of people of color.

Famously, a courageous educator named Jane Elliot did just that in a classroom in Iowa in 1968. In the wake of the shooting of Dr. Martin Luther King Jr., Elliot designed an exercise to teach the children of her third-grade classroom about racial discrimination. She separated the class of children into those with blue eyes and those with brown or green eyes. She then instructed the children with brown eyes to wear construction-paper armbands, and told the children that brown-eyed and green-eyed children were smarter and cleaner than blue-eyed ones. The experiment lasted a few days, and the roles were switched partway through; blue-eyed children became the dominant, smart ones, and the brown and green became less smart and clean.

Elliot made it onto Johnny Carson from the tiny burg of Riceville, Iowa, because of her exercise, which gained national attention, praise, and scorn. Some decried Elliot for subjecting the children to such a cruel experience. Some said she was a pioneer in civil rights. She's frequently noted as one of the most influential educators of the twentieth century thanks to her eye-color exercise.

The lesson has been reframed and retaught across the world in various incarnations, but from an embodied view, it makes sense. Say what you will about the cruel way it turns children against each other (although spend a few minutes on the playground, or around the lunch table of middle schoolers, and your frame of reference may shift), but there's no doubt that this embodied process of teaching racial discrimination is a powerful methodology. It calls on the lived, sensorimotor experience of the participant. No PowerPoint can do that.

THE WHITE MAN'S EMBODIMENT

Rudyard Kipling infamous poem, "The White Man's Burden," is a racist screed that suggests that it is the role of the white man (read here, Western European and American men) to civilize and instruct the other races of the world, essentially anyone with brown or black skin. Kipling calls the people of the world who aren't white "sullen peoples/half devil and half child." Obviously, this colonialist viewpoint has been razed by professors in universities around the world as an example of the kind of odious imperialism that subjugated so many of the world's people.

And yet in many of those same universities, conditions still exist in which racism abounds. There are different forms, usually, such as microaggressions and white hires, tokenism and condescension. But Kipling had one thing right: the burden belongs to white people.

As the research is beginning to make abundantly clear, humans opt in to groups like themselves based on skin color. The only way to mitigate the effect of this "in-crowding" based on race is to empathize with people who have other skin tones by way of embodiment. White institutions — including schools — would do well to pay heed to the notion that in order to offer empathy and inclusive gestures to members of other races, white individuals are going to have to embody other races. How this is to be done is tricky. However, if diversity, equity, and inclusion efforts are truly going to take hold, white institutions (most institutions in the United States are "white" in the sense that they adhere to Western European values, rules, cultural forms, and ideologies) are going to have to address the idea of "whiteness" and how that very embodied notion impedes white individuals' ability to perceive other races from an embodied perspective.

Whiteness as an embodied construct can be summed up in the classic little parable of the fish, famously paraphrased by the late writer David Foster Wallace at a Kenyon College commencement speech. It goes something like this:

> Two fish are swimming along, minding their own business. An older fish comes swimming by them in the opposite direction, and says, "How's the water, boys?" The two fish continue swimming, and after a moment one says to the other, "What the hell is water?"[11]

The joke is that we can never see what's all around us, all the time, and the most obvious realities are sometimes the hardest to perceive. That's what whiteness is like for most white people; perception is so subjectively based in the individual experience of being white that it's very hard to see race through a lens other than white.

This is where embodiment comes into play. Only through empathy — an embodied response to the pain of another individual from what is perceived as the "like" group — can schools and education create a way forward that addresses whiteness. People of color have had the burden of educating about race for too long. It's time for the white man to take up the burden.

NOTES

1. Sandhya Somashekhar, "The Disturbing Reason Some African American Patients May be Undertreated for Pain," *Washington Post*, April 4, 2016. Retrieved from https://www.washingtonpost.com/news/to-your-health/wp/2016/04/04/do-blacks-feel-less-pain-than-whites-their-doctors-may-think-so/.

2. Albert Newen, "The Embodied Self, the Pattern Theory of Self, and the Predictive Mind." *Frontiers in Psychology* (Nov. 23, 2018). https://www.ncbi.nlm.nih.gov/pmc/articles/PMC6265368/.

3. Matteo Forgiarini, et al. "Racism and the Empathy for Pain on Our Skin." *Frontiers in Psychology* (May 23, 2011). https://www.ncbi.nlm.nih.gov/pmc/articles/PMC3108582/.

4. Tabitha Peck, et al. "Putting Yourself in the Skin of a Black Avatar Reduces Implicit Racial Bias." *Consciousness and Cognition* (Sept. 2013). https://www.ncbi.nlm.nih.gov/pubmed/23727712.

5. Alexandra Kalev, Erin Kelley, and Frank Dubbin, "Diversity Management in Corporate America" *Contexts*, vol. 6, no. 4 (2007), 21–27.

6. Kalen, Kelly, and Dubbin, "Diversity Management in Corporate America," 25.

7. Kalen, Kelly, and Dubbin, "Diversity Management in Corporate America," 22.

8. D. R. Godden, and A. D. Baddeley. "Context-Dependent Memory in Two Natural Environments: On Land and Underwater." *British Journal of Psychology*, vol. 66, no. 3 (1975), pp. 326. doi:10.1111/j.2044-8295.1975.tb01468.x.

9. Godden and Baddeley, "Context-Dependent Memory in Two Natural Environments: On Land and Underwater," 326.

10. "Watching Others Makes People Overconfident in Their Own Abilities." *Association for Psychological Science (APS)*. https://www.psychologicalscience.org/news/releases/watching-others-makes-people-overconfident-in-their-own-abilities.html.

11. David Foster Wallace, "This Is Water," Kenyon College Commencement Speech, May 21, 2005. Retrieved from https://fs.blog/2012/04/david-foster-wallace-this-is-water/.

THIRTEEN
Anything Boys Can Do

Most educators and teachers make efforts, at least on the surface, to treat male and female students equally. Thanks to increased tolerance and education around issues of sexuality and gender, many schools now have a growing population of students who identify as gender fluid or gender nonbinary or transgender, and they are increasing local and national efforts to treat those students in a respectful, appropriate manner as well. This isn't to say the battle for acknowledgment of gender diversity has been won—far from it. But leaps and bounds have been made in the past thirty years.

However, treating students the same, and not taking into account their contextual reality, ignores the embodied nature of subjective experience. It is this subjective, sensorimotor reality—which is itself tied very much to gender—that provides students with their sense of self. Treating students "the same" and patently ignoring race and gender, among other characteristics, isn't always a particularly helpful approach. Cisgender women, for example, experience a walk home through darkened streets late at night differently from a two-hundred-pound white male. One cannot equate the experience when the lived sensorimotor experience is so different. To treat genders equally but also with respect to individual identity is tricky and requires a concerted effort and education around gender differences.

As a starting point, it's helpful to acknowledge that—just as in implicit bias—most people share a set of beliefs and understandings about men and women, whether they are borne out in evidence or not. In many ways, this echoes the pioneering work of George Lakoff and Mark Johnson, who argue that language is largely embodied through metaphor.[1] For example, participants in a study judged issues as more important when the clipboard they happened to be holding when reading about the issues was heavier. Weight—a sensorimotor perception—equals seriousness.

As another example, if you visit two dorm rooms, one filthy with nasty old ramen noodles on the floor and the other clean with a nice lavender candle, you'll judge the inhabitants of the cleaner room as more morally pure. Humans associate context with feelings. The object of sensorimotor perception changes the manner in which that object is classified, cognitively speaking. Thus, when we are confronted with two individuals, one a cisgendered male and the other a cisgendered female, those contextual, perceived cues affect the way we think.

FLUFFY BALLS!

American culture has long associated women with softness and men with toughness. To bring this point home, researchers from Tufts University had participants look at a bunch of digitally altered faces that were gender ambiguous—they were computer designed to be exactly in the middle of male and female.[2] When folks in the study held a hard, tough ball when looking at the pictures, they judged more faces as male. A soft, fluffy ball? More faces were identified as female. This suggests, according to the authors, that ideas about gender are embodied, despite our abstract attempts at gender equality. As the researchers put it, "social category knowledge is at least partially embodied."[3]

The research suggests that it's likely when individuals engage with different sexes they have embodied ideas about them based on their difference; they don't treat everyone the same. This is seems like kind of a bland statement that seems a bit obvious but is in reality a pernicious situation. If educators perceive female students as less tough than their male counterparts, do they challenge them as much in school? What about in sports? Perhaps faculty and staff don't invite certain women to contentious meetings because of a subconscious belief that they're not "tough" enough to handle it.

Toughness can be not only an emotional quality but an intellectual one as well. Do educators—whether professors in higher education or kindergarten teachers—inherently ask less of female students, not realizing that they believe them not to be up to a difficult intellectual task? Men usually get more attention in educational settings, although that's changing. White male graduate school applicants are more likely to be granted interviews with advisers than women or people of color, for example.[4] Again, this strongly suggests a situation that hints at interoceptive abilities. If beliefs about women being "tender" are embodied, then a reasonable approach to counteracting that embodied bias is by developing an ability to be more in touch with interoceptive states.

Although most folks in education would state that they tacitly understand the bias against women as not being "tough" is unfair, they may feel on some level there is a difference in the emotional life of men and women. Women are tough, no doubt—watch Serena Williams play tennis or take a look at Ruth Bader Ginsburg's career—but it does seem that treating every single student the same isn't a very nuanced approach that takes into account personal, individualized, subjective experiences and its relationship to behavior. There are, it seems, some ways in which men and women differ in a fashion that directly affects their learning: reading comprehension.

Here's the scenario: homework is done, kitchen cleaned. A brother and sister open a box of Chips Ahoy! and plop on the couch, ready to veg out and watch a movie. The interminable scroll through Netflix and Amazon begins: what to watch? The sister's choices are inevitably independent dramas of some kind. She cruises right past *Thor: Ragnarok* and *Guardians of the Galaxy II* and picks movies about unrequited love, family discord, and melancholy landscapes. Sad stuff, basically. It's what the kids ends up watching because the brother gives in, despite his desire to watch another John Wick movie.

Remember that Lakoff and Johnson note that language, particularly metaphor, is embodied.[5] Language comprehension, broadly construed, is as well. Readers simulate many of the feelings they read about. Arthur Glenberg of Arizona State University puts it like this: "Language is understood through a simulation process that calls upon neural systems used in perception, action, and emotion. Thus, simulating the language produces much the same effect as being in the situation."[6]

To put it simply, when you write, "She dropped the large rock on her toe, squashing it and smashing the toenail," the parts of your brain and nervous system that would respond if your toe was *actually* smashed respond. We experience the lives of the characters on the page more viscerally than we might realize. That's why the scene in Stephen King's *Misery* wherein Annie Wilkes, a crazed super-fan, cuts off the foot of Paul Sheldon and cauterizes the bleeding stump with a blowtorch is so gruesome to read. It makes one's ankles tingle just thinking about it.

To illuminate the influence of gender and reading comprehension, Glenberg's research is quoted at length. Glenberg and his coauthors list five predictions based on their research:

> 1. Being in an emotional state congruent with sentence content facilitates sentence comprehension.
>
> 2. Because women are more reactive to sad events and men are more reactive to angry events, women understand sentences about sad events with greater facility than men, and men understand sentences about angry events with greater facility than women.
>
> 3. Because it takes time to shift from one emotion to another, reading a sad sentence slows the reading of a happy sentence more for women than men, whereas reading an angry sentence slows the reading of a happy sentence more for men than for women.
>
> 4. Because sad states motivate affiliative actions and angry states motivate aggressive action, gender and emotional content of sentences interact with the response mode.
>
> 5. Because emotion simulation requires particular action systems, adapting those action systems will affect comprehension of sentences with emotional content congruent with the adapted action system. These results have implications for the study of language, emotion, and gender differences.[7]

This is pretty profound stuff; it's slightly odd, but it also makes sense. If you're sad and read something sad, *you just get it, man*, in a way that someone obliviously happy-go-lucky won't. There's clearly a recursive quality here: angry people seek out angry writing because it resonates with them and probably feeds their anger more. It makes the appeal of shows that market anger and hate—Alex Jones's *Infowars*, Rush Limbaugh's radio show, Michael Savage—easy to understand. They all traffic in a particular brand of hateful rhetoric, which is easily understood by their listeners who share this anger-based worldview. This ability to more easily comprehend emotions when in the emotional state aligned with the reading extends to various other facets of embodied comprehension, including body language.

Here's the scenario: a teacher is giving a lesson to a mixed group of students, both male and female. (It's important to note before we go further that some students identify as gender nonbinary, and thus are not accounted for in the following scene.) The teacher does her thing, and while teaching the lesson goes through various stages of emotion, either excitedly describing how cool a recent study is, or gets a bit angry and worked up about the human rights abuses in Yemen. The question is, are both male and female students on equal footing in terms of comprehension?

According to Glenberg and colleagues, clearly reading comprehension levels differ depending on the text being read.[8] Already we have one difference; women comprehend sad information more clearly and men understand anger-based text more.

In addition, the ability to comprehend what's being read also depends on the *mood* of the reader. If you're already in a nasty mood because you forgot to buy coffee and got an overdue bill notice that morning, and you read a particularly vituperative screed about the Russian interference in the 2016 election, you'll huff and you'll puff and be able to spit direct quotes from that article all day. Conversely, if you slept in, had coffee and a delicious bagel sandwich with your partner, and received a random check in the mail because you accidentally overpaid a bill, you won't parse the meaning of the Russian-interference article in quite the same way.

The teacher continues her lesson, and the students absorb the information based on sex and mood at different rates and with differing accuracy. Not only that, but the professor or teacher is bouncing around the room (as many educators do, trying to compete for limited attention spans) trying to get the lesson across. She's throwing her whole body into this one. And here's where embodiment again plays a crucial role. According to some pretty wild and edgy research from Germany, the gender of the student will affect their ability to read body language.[9]

The study in question measured the difference between males and females in interpreting body language. Males, it turns out, surpass women in recognizing happy body language; women surpass men at identifying angry body language. In addition, women were much more attuned to picking up neutral body actions.

The authors of the study—all of whom hail from Eberhard Karls University of Tübingen—go on to discuss some interesting possible ramifications of these findings, paraphrased here.

The research suggests men are more skilled at detecting subtleties in happy faces than women. This stands in direct contrast with some widely held beliefs about the gender differences and the ability to read emotions. Women are usually credited with a more astute perception of emotion, and the reason is often attributed to their role as mothers and caregivers. Complex social interactions and complicated emotions are usually seen as the purview of women, whereas men's social cognition is frequently assumed to deal more with reactions to situations; active responses versus more delicately calibrated emotional analysis. The research seems to suggest, however, that men are more skilled at detecting happiness than women, and women are better at detecting emotional neutrality. It's not exactly a reversal of conventional wisdom, but certainly different.

Men's and women's brains respond differently to happy faces. When stuck in a functional magnetic resonance imaging (fMRI) machine, the amygdalas of men and women respond differently to pictures of happy faces. This suggests that the "social brain," a network of neural structures that includes the amygdala, is different between the sexes. When men are shown fearful faces or aggressive faces, their amygdala response is different as well; this makes sense, as those situation-dependent responses are tightly tied to testosterone levels.

We don't know if there is a neurological underpinning for how different sexes read body language. Although there's bushels of research in which men and women have been stuck inside an fMRI machine and had their brains scanned while looking at pictures, not a whole lot has been done with body language. There's also another sociocultural layer here. Context matters, and the environment we're in and the way we learned social behaviors based on our cultural history can change the scenario.

Back to the classroom, where our sweaty, underpaid teacher is struggling to get the lesson across, keep the kids from checking Snapchat on their phones, and get the salient points of the lesson across. What we can say with a least a decent measure of surety is this: because experience is subjective, and subjective experience is embodied in the sensorimotor system, and that system is interconnected with many other networks in the body, including the endocrine system, it's relatively reliable that men and women are experiencing the lesson, emotionally and cognitively, differently. To a certain extent.

BODY MATTERS

This isn't all that earth shattering—there are so many qualifiers that it almost seems inconsequential. However, if teachers start thinking about the basic structures of learning within the modern educational system—reading comprehension, for example—it does seem possible that the embodied experience of men and women differ. Can educators reasonably expect, then, for different genders to have the same experience in school, and be assessed the same way, with the same criteria? Grades have long been seen as poor indicators of performance, and even more faulty if the extra contextual layer of gender is laid over assessment. It's vital to acknowledge and understand the role sex plays in the classroom, insofar as that's possible.

The way a man and a woman—and for that matter, an individual who is gender nonbinary, transgender, or otherwise doesn't fit the traditional division of the sexes—experiences education may be markedly different. The body matters.

Luckily, the research that has developed to look at the difference in the sexes from both an embodied and neurological view is rich.[10] It can help answer a tricky question: are men and women different in some way when it comes to learning? It's widely understood that cognitively they are capable of the same feats; whether it's science or literature or art, both sexes are capable of miraculous feats of invention. But there are differences that bear on education, particularly in the way that brain structure relates to our hominid evolution. Others, including the structural difference of various parts of the brain, also relate to various aspects of learning.

In review of contemporary research on sex differences in the brain published in *Nature*, author Larry Cahill helpfully catalogs some of the differences that exist in male and female brains, and the way the sexes may embody learning differently.[11]

During the past five million years or so, males and females of the hominid line—beginning way back with ancestors such *Australopithecus aarensis* and leading up George Washington, Hedy Lamarr, Kanye West, and Jennifer Lawrence—had lives that faced many selective pressures. Some of those selective pressures, meaning the factors that drive evolutionary adaptation like the way our brains work and how we learn, were the same for both sexes, some different.

For example, researcher Kazuhito Tomizawa reported that the hormone oxytocin, which regulates lactation and other functions related to pregnancy and birth, also increases spatial cognition, particularly memory.[12] This makes sense, as when a mother has a bunch of babies running around, it pays to have a good memory for where things are. It's a conserved adaptation for a mother to know all the little nooks and crannies and spots when she's trying to wrangle her children.

Now, luckily, we have the magnetic tractor beam of Nintendo for that. In addition, there's an argument to made—related to the points regarding body language earlier—that the subtle ways females of the hominid line have maximized social strategies have a direct effect on their ability to read more nuanced social cues. Both of those scenarios relate to various parts of the brain, which do, in fact, show structural differences between the sexes, particularly the hippocampus, which is designed for memory (among other functions), and the amygdala. This bears out in the research; women generally fare better at fine-detail memory than men.

There is a wealth of information, some of it contradictory, and all of it complex. There are no easy answers when looking at the embodied nature of learning through the frame of gender and sex.

Here's the basic take away from all this. It appears, through the lens of neuroscience at the very least, that there are measurable differences not just in the brains of men and women, but in the resulting cognitive abilities. Given what is known of embodied cognition, the very prism through which subjective experience is felt is the body; thus, it is "gendered" in the sense that perception of the world is collected by embodied senses. Perhaps the easiest way to think about it is as follows: women and men experience the world according to their biological sex and also how that sex is defined by our society. But in both cases, the body receives and processes that information, be it from interoceptive sense or from sensorimotor interaction with the world, and that information is part of embodied meaning making.

The history of treating women as inferior is toxic. Unfortunately, it still happens, even in the most progressive institutions. For another example, watch any televised debate between male and female political candidates and you can start to tease out the way women get pigeonholed unfairly. No teacher should teach differently to different genders—we don't want to go back to the times when women did needlework and men studied philosophy. Certainly, however, we should be nuanced in our appreciation of how male and female students, respectively, interpret text, situations, and ourselves.

NOTES

1. George Lakoff and Mark Johnson. *Philosophy in the Flesh: The Embodied Mind and Its Challenge to Western Thought*. New York: Basic, 1999.
2. Michael L. Slepian, et al. "Tough and Tender." *Psychological Science* vol. 22, no. 1 (2010), 26–28. doi:10.1177/0956797610390388.
3. Slepian, "Tough and Tender," 26.
4. Mary Romero, "Reflections on 'The Department Is Very Male, Very White, Very Old, and Very Conservative': The Functioning of the Hidden Curriculum in Graduate Sociology Departments." *OUP Academic* (Apr. 8, 2017). https://academic.oup.com/socpro/article/64/2/212/3231922.
5. Lakoff and Johnson, *Philosophy in the Flesh*.
6. Arthur M. Glenberg et al., "Gender, Emotion, and the Embodiment of Language Comprehension," *Emotion Review* 1, no. 2 (2009): 151–61. doi:10.1177/1754073908100440.
7. Glenberg et al., "Gender, Emotion," 151.
8. Glenberg et al., "Gender, Emotion."
9. Arseny A. Sokolov, et al. "Gender Affects Body Language Reading." *Frontiers in Psychology*, vol. 2 (Feb. 2, 2011). doi:10.3389/fpsyg.2011.00016.
10. Larry Cahill, "Why Sex Matters for Neuroscience," *Nature News* (May 10, 2006). www.nature.com/articles/nrn1909.
11. Cahill, "Why Sex Matters for Neuroscience."
12. Kazuhito Tomizawa, et al. "Oxytocin Improves Long-Lasting Spatial Memory during Motherhood through MAP Kinase Cascade." *Nature Neuroscience*, vol. 6, no. 4 (2003), 384–90. doi:10.1038/nn1023.

FOURTEEN
Golden Oldies

Of all the indignities of getting old, perhaps none are quite as galling as the way the body begins to *feel*. The outward appearance of the aging body is more easily accepted—after all, surrounded by individuals at various stages of their lives, the fact that the body begins to decline is evidenced daily by the various people one interacts with. And although age can often bring benefits, financial security, wisdom, a prominent role in social groups, hierarchical standing, early-bird specials at Denny's, and so on, there is little doubt that as the body ages there is simply a decline in *feeling good*. Joints stiffen and ache. Muscles atrophy, and vision, hearing, even taste can become less acute. For some, sexual desire decreases. Older people simply suffer from more glitches and faulty wiring.

Watching old men try to raise themselves out of plush recliners is nothing but a sad reminder that the body is destined to break down, and the physical limits imposed by aging and the concurrent inability to engage in a life fully embodied is the price human beings pay for hanging around too long. Over time, the body falters.

But does it have to?

Emerging science and your own, subjective experience can shed some light on this question. From an embodied cognition perspective, if the body is aging—that is, no longer renewing itself on a cellular level with vigor as it did during earlier stages of development—then the mind, too, is in decline. Again, however, it's important to note here that if the central tenets of embodied cognition are true, then the mind/body (or embodied mind) is in decline. Age deteriorates the whole kit and caboodle.

People experience mental lapses more as they get older, or so it would seem. Anyone who has spent more than four decades on this planet knows the pain of misplacing car keys, or the bewilderment of walking upstairs only to discover that you can't for the life of you figure out why you ascended in the first place. Add the attention-shattering effects of the internet and smartphones to the equation, and most folks should consider themselves lucky that individuals near a half century old and older don't walk around pantsless and unwashed on an endless quest to remember where they put their wallet.

The embodied mind—cognitive, physical abilities, emotions, the whole shebang—would, seemingly, go down the tubes all at once. As noted, one of the theoretical benefits of getting older should be wisdom. As individuals lose touch with their physical selves and slow down, get creaky, start griping about back pain, and performing a symphony of grunts when extricating themselves from a low-slung sport sedan, there should be a rise in wisdom, an increase in fluid intelligence measures.

Fluid intelligence is a concept that was developed by Raymond Cattell in the late twentieth century to describe problem solving, the ability to negotiate relationships, and abstract reasoning.[1] This type of thinking is different, in that it is used in situations in flux, and draws on historical understanding as well as current situations to draw conclusions or accurate speculations without clear evidence. In a nutshell, it's the most nuanced and complex form of thinking—at least as far as many traditional cognitive scientists are concerned.

Matthew Costello is an experimental psychologist at the University of Hartford. Some of Costello's research addresses the way in which individuals may become less embodied over time.[2] When I reached him by phone, Costello generously walked me through some of the surprising twists and turns of the effect that aging can have on embodiment.

One of the key factors in assessing the effects of aging on embodiment is multisensory input (MSI). Humans receive sensory input all day long—visual, auditory, haptic, olfactory, but also the less noted senses such as proprioception and interoception, and even balance as sensed through the vestibular system. MSI is the way humans integrate all this incoming stimuli into a picture of the world and our place in it. Costello has found that it appears likely that over time older adults favor their *visual* perception more and more over other senses. This shift away from multiple means of collecting information from the environment brings with it a shift in embodiment.

"There's a link between knowledge and sight," Costello pointed out. He was referring both to current research on the way older adults favor visual perception but also a long history within philosophy of equating visual understanding with cognitive processes, hence the phrase "seeing is believing" rather than "smelling is believing." This second phrase does become socially codified, however, for parents once their kids hit puberty.

Essentially, Costello has noted a decline in embodiment along with the prevalence of visual perception as the primary modality for gathering information about the environment in older adults. In essence, the body grows less attuned. But *why* this shift occurs isn't clear. When older adults, aged between approximately fifty and eighty, lose their sense of embodiment, is it simply because they're aging? Or is it because older adults adopt a lifestyle that inhibits the fine-tuning of their embodied selves? "Maybe at some level from a psychological perspective they shift away from being embodied," Costello said.

Although getting older is the most natural thing in the world, there are some parts of aging that we don't understand particularly well. Aging "is a normal process that would occur, but it's not that it's inevitable," according to Costello. He noted this to me as he described the manner in which aging exhibits itself in older adults. Although this seemed counterintuitive—of course we age, we're designed to—there was a chicken-and-egg scenario at work, which Costello explained to me. Do we become less embodied because we age or do we age because we become less embodied?

To work toward an answer—or simply to better understand the question—it's beneficial to define and contextualize what *embodied cognition* means in terms of how it shapes interactions with our environment. Researcher Massimiliano Cappuccio states it as follows in the journal *Phenomenology and the Cognitive Sciences*:

> It is the history of the recurrent interactions between body, environment, and social context that defines sophisticated patterns of sensorimotor correspondence, defining at once a worldly horizon of viable opportunities for action and a correlated embodied "self" that is ready to take advantage of these opportunities.[3]

Humans' ability to cognitively act is predicated on our history of physical action. The experiences people have, bodily, in the past are drawn upon to develop a sense of self as well as inform decisions and abstract thinking and emotions. It follows that a decrease in "interactions" of the kind suggested by Cappuccio would lead to a decrease in the ability to cognitively benefit from past experiences.

Consider an example scenario: in your youth, despite the emergence of a digitally dominated culture, you're banging around quite a bit; jostling with classmates at school, playing at recess, climbing trees, having adventures, arguing with siblings. In your younger years, you were simply *out* more, moving physically and emotionally and socially. Going to college or getting jobs, maybe starting a family come after, and it's possible your twenties and thirties are filled with the sort of experiences that lead to "worldly horizons."

Many jobs individuals have when young are practically bursting at the seams from an embodied perspective. Early occupations include working in orchards, dishwashing, and landscaping; many folks wait tables or work retail, all jobs that have some degree of embodied action, even if it's only folding a pair of khakis at the Gap. Life is a cacophony of sensorimotor experiences during your early years.

Eventually, however, and hopefully, a stable job is found. No more shoveling or cleaning or moving about for you; you move into the office. You sit. You become a middle manager. Your children grow older, and there's no need to run about chasing them here and there any longer. You become increasingly sedentary. Few individuals play sports anymore compared with their adolescence, when expressing yourselves bodily through athletics (and other less socially sanctioned bodily pursuits) was second nature. You simply dial back on the sensory input maelstrom of your youth and rely primarily on visual perception versus the MSI of your earlier years. The shift from embodied youth to disembodied old age is in full swing. Pretty soon you'll have to put those adhesive gripping stickers on the bathtub floor (if they're not there already) to avoid a slip.

Whether or not a change in lifestyle from an active, embodied youth to a more sedentary, disembodied middle age causes older adults to be less embodied is unclear. What does seem easier to prove is that *our sense of embodiment is plastic and variable and able to be changed.*

"Older adults would benefit from multisensory input," Costello noted. The ideas that some regimen of sensory input for older adults would be beneficial in slowing or reversing some of the typical symptoms of aging isn't necessarily that far-fetched. Multiple articles abound that espouse the benefits of crossword puzzles and the like for older adults to aid in keeping the mind and memory sharp. But if the mind is embodied, crosswords alone won't provide the same benefit as multisensory experiences. Older adults who engage in all manner of sensory conditioning often have a correlated cognitive improvement as well. "As the body gets more tuned up we could expect to see some changes" in cognition, Costello suggested.

LIFELONG LEARNING AND THE EMBODIED SELF

The idea of "lifelong learning" permeates education to an impressive degree. Yale University has an office of lifelong learning attached to their divinity school that provides offerings for alumni to continue their education. The Princeton adult school states that "Learning Never Ends," and the office of the vice provost at Stanford is committed to ensuring a "lifetime of learning opportunities." Even the US Department of Education is in on the game, calling lifelong learning "A Roadway to Success."

Theoretically, it's understood that adults can learn as they grow older. A 2017 article in *Nature* titled "Age Is No Barrier: Predictors of Academic Success in Older Learners" says as much.[4] But how much humans are able to learn—and really, whether they really do at all—is also related to a number of factors.

Do adults stop learning as much because a life of sick kids, grocery shopping, home repair, long hours at the office, and streaming Hulu shows simply sucks up too much time, or is there some neurological mechanism that makes it more of a challenge? Is it environmental, situational, or some idiosyncratic personality trait that keeps some people hungry for intellectual development late in life? The *Nature* article makes one thing clear: adults can learn. It states, "Ageing does not impede academic achievement, and . . . discrete cognitive skills as well as lifetime engagement in cognitively stimulating activities can promote academic success in older adults."[5]

Researcher Willis Overton makes this claim in regard to the nature of embodiment and experience:

> Embodiment is the claim that perception, thinking, feelings, desires—the way we behave, experience, and live the world—is contextualized by our being active agents with the particular kind of body we possess. In other words, the kind of body we have is a necessary precondition for having the kind of behaviors, experiences, and meanings that we have.[6]

Consider the phrase "particular kind of body we possess." This suggests that the relationship to the body, particularly when folks get older, bears directly on the way in which experience connects to cognition and development. Therefore, if educational institutions promote the idea of creating "lifelong learners," it seems reasonable that attention should be paid to implementing and nurturing an embodied approach to learning throughout the early developmental stages.

There is an incremental but inevitable decline in the way humans react and respond to the environment and stimuli. In earlier developmental states, children's sensorimotor engagement with the environment is visible. Toddlers, in particular, are always tasting the world around them, touching things, picking up various items—a full-on sensorimotor exploration of the world. By the time kids hit adolescence, they've transferred to largely sensory-antiseptic classrooms, where tactile experience is eschewed in favor of pure cognitive tasks.

But the manner in which sensorimotor systems—taste, sight, touch, smell, sound—interact with neurology and behavior haven't suddenly changed. What happens is that there is a prioritization of the visual over every other sense, and abstract cognitive functions over more embodied processes. As adults migrate to cubicles and spreadsheets, the full-body, sensorimotor appreciation of the world and its resultant increase in the neural pathways atrophies.

The question educators need to ask of one another and themselves—or anyone interested in maintaining a love of learning and growth and exploration over the course of his or her life—is how can one nurture a life of learning that fully engages the sensorimotor systems in their entirety over the course of a life? The answer lies in daily life, in the miraculous way that bodies interact with their environments. By engaging in mindful appreciation of the body's reciprocal relationship with the world, people become more attuned to that environment, and build those networks that support learning and memory. Take a hike in the woods or explore the verge next to the storm-water pond in a suburban development. Any exploration that encourages the full sensory appreciation of the world engages those systems and keeps them functioning and viable pathways for learning.

The alternative isn't all that appealing: a slow decline in cognitive powers, atrophy of cognitive skills, and the decay of physical abilities and the senses. In a 2018 *New York Times Magazine* article profiling septuagenarian cross-Atlantic Polish kayaker Aleksander Doba, the adventurer repeats a well-known Polish saying when asked why someone his age would risk loneliness, injury, and death crossing the Atlantic (multiple times!) in a fragile kayak on a dangerous journey that takes weeks, even months: "I do not want to be a little gray man," Doba says.[7] His sentiment is shared by ninety-two-year-old Australian speed walker Heather Lee, who holds multiple world records. When asked in another *New York Times* article if she'd like to die walking, Lee responds "Wouldn't that be lovely?"[8]

These are extreme examples, to be sure. But certainly the spirit of what Doba and Lee project is one that could be harnessed in lifelong education efforts. Movement and constant engagement of the senses is one way to ensure that vital neural networks stay up to snuff.

As students transition from childhood to adolescence, young adulthood to midlife, the importance of movement, of exploration, becomes paramount. A great example is the little old man in the animated film *Up*, who is thrust into action as an aged protagonist, and finds himself worthy of the challenge. If the idea of learning was synonymous with movement, with sensorimotor engagement, and lifelong learning was embraced by many, one wonders what the long-term effects would be. Many adult learners—many adults, for that matter—share the Polish kayaker Doba's fear. And the only way to combat that—for teachers and students—is to constantly reinforce the idea of full-body engagement with the world as the path to true learning.

NOTES

1. Richard E. Brown, "Hebb and Cattell: The Genesis of the Theory of Fluid and Crystallized Intelligence." *Frontiers in Human Neuroscience*, vol. 10 (2016). doi:10.3389/fnhum.2016.00606.

2. Matthew Costello, personal interview.

3. Massimiliano L. Cappuccio, "Introduction: When Embodied Cognition and Sport Psychology Team-Up." *Phenomenology and the Cognitive Sciences*, vol. 14, no. 2 (2015), 213–225. doi:10.1007/s11097-015-9415-1.

4. Abbie-Rose Imlach, et al., "Age Is No Barrier: Predictors of Academic Success in Older Learners," *Nature News* (Nov. 15, 2017). www.nature.com/articles/s41539-017-0014-5.

5. Imlach, et al. "Age Is No Barrier."

6. Willis Overton, et al. "Developmental Perspectives on Embodiment and Consciousness," *Psychology* (2013), 18.

7. Elizabeth Weil, "Why He Kayaked across the Atlantic at 70 (for the Third Time)," *New York Times* (March 22, 2018). www.nytimes.com/interactive/2018/03/22/magazine/voyages-kayaking-across-ocean-at-70.html.

8. Bari Weiss, "Try to Keep Up with Australia's Fastest 92-Year-Old Woman," *New York Times* (Jan. 4, 2019). www.nytimes.com/2019/01/04/opinion/australia-racewalking-aging-healthy.html.

FIFTEEN
Dionysian Embodiment

He'd always been a troublemaker. Wild. Ungovernable. Unpredictable. Complicated in that his power was great, yet wasted on frivolity, on song and dance and revelry. His home was tangled and overgrown, rank with weeds and shoots. His friends were half-beasts. Dionysus was chaos incarnate—a symbol of confusion and complexity.

To the ancient Greeks, Apollo was a more fitting god. Plato and Socrates, particularly, were fans. Apollo was a god of order, symmetry, division, and compartmentalization, a symbol of organization and reason, of anal-retentive middle management. Nietzsche would write about this contrasting duality, using an analysis of Dionysus and Apollo and the way they defined Greek tragedy to delve into ideas of order versus disorder, rationale versus chaos, the world of defined ideas versus the abyss.

Many cultural concepts and institutions—art, medicine, government, philosophy, and economics—can be traced back to the ancient Greeks. Conceptual frameworks, such as ideas of order and disorder, also have roots in the classical tradition. In particular, one of the institutions that most clearly rests on the Grecian traditions is education. And within education, both that inherited from the Greeks but also many of the inherent structures developed since ancient times, there is a prevailing emphasis on rigor, order, delineation, and compartmentalization—the stuff of Apollo's rule.

Looking at learning through a more Dionysian lens—the messy, interconnected and intersubjective, manner in which the body plays a role in development and cognition—offers a richer and ultimately more rewarding approach to the project of human education. It is not well packaged, to be sure. If this book suggests anything, it is that the relationships among behavior, cognition, memory, emotion, the brain, the senses, and the body are incredibly complex. But to leave the body out of the equation is an impoverished way to approach the project of teaching and learning. Dionysus may have been wild and tangled and unpredictable (he reminds me of some of my more eclectic colleagues' hair) but the life of learning led by his example is one in which all facets of experience, including the sensorimotor perception of the world, are taken into account.

Learning from an embodied perspective is ultimately more influential, more meaningful, and longer lasting.

Apollo and Dionysus are also helpful allegorical characters to use to aid in unravelling the messy knot of how humans think. Using these two symbolic characters as a springboard, one can ask big questions that bear on education. What is the mind? What is consciousness? And how is it that humans can be self-aware, imagine other worlds, and use their intellect to wield tools, communicate, and build complex societies? And, perhaps more to the point, what is changing within education—for good or ill—in relation to how students learn and think within a rapidly changing world immersed more and more in indoor, screen-based activities?

RECLAIMING THE INTERSTITIALITY OF CONTEMPLATION AND WONDER

It seems clear that within the modernized, digitized, industrialized world, individuals are becoming more and more divorced from direct engagement with corporeal reality. Whereas people once relied on complex visuospatial abilities to navigate the world—a messy combination of sensorimotor input, memory, emotion, and neurochemistry—humans now rely on Google Maps to get places. The clocks no longer require winding, the cars drive themselves. Automation is here. In *The Glass Cage*, Nicholas Carr brilliantly documents these changes.[1] The tsunami of intermediating digital influence has already swamped us.

The interstitial moments when humans used to be in their bodies—walking down the street, only to veer off and take the shortcut under the high-tension wires; the hunt for a book to find a quote they've been trying to pin down; the aimless room-to-room wander anyone who grew up prior to the web remembers from long, boring, unstructured days—are now pockmarked with glowing digital portals. All of us walk with phones in hand, scrolling. Or, worse, we take in our surroundings through the anticipatory lens of social media feeds. What will make a good post? (I find myself composing sentences in the form of funny, off-handed tweets.)

As a kid I was obsessed with Marvel comics, using my hard-won lawn-mowing money to buy *X-Men* comics. I loved the dark-skinned Nightcrawler character, with his acrobatic circus roots and teleportation abilities. To embody the love I had for this mutant hero, who only had two fingers and a thumb instead of the usual four-to-one of a human hand, I'd tape my pointer and middle fingers together, do the same for my pinkie and ring fingers, and run around pretending to be Nightcrawler, even making the classic "BAMF!" sound that announced his teleportation. In love with the daredevil heroics of Evel Knievel, I'd build jumps in my driveway made of old stove wood and planks in an effort to re-create his epic launches with my BMX bike.

But what use are those methods of embodied childhood hero worship now, when video games offer up a hyperreality so much more pungent and vitiating, so perfectly designed to tap into the addictive dopamine feed in the adolescent brain? An eleven-year-old running around with scotch-taped fingers seems a relic of the past. The ambrosia of slick digital experiences is compelling; this isn't news. They beckon, and we obey.

Education has followed suit, further relegating the messy, hackneyed, embodied experience to the occasional art room, sports field, and drama club. Recent cuts of government funding for groups such as the National Endowment for the Arts, the National Endowment for the Humanities, and the Corporation for National and Community Service underscore the contemporary indifference toward embodied learning—these programs often fund after-school arts and music activities across the nation.

What's favored instead is the austere and categorical accumulation of information and skill sets, algorithmically articulated advances in career potentialities. Libraries become less focused on housing books, more concerned with acting as "hubs" of "information literacy." Those fighting to keep an embodied approach to learning like Montessori schools or private schools like Whitby in Greenwich, Connecticut (both of which charge tuition, thus relegating embodied education to those with wealth and furthering an unequal society), still send students off at the end of the day to homes with Wi-Fi and an endless stream of digital options for entertainment and absorption.

I walk into my classes here in New England, and you could hear a pin drop. Students are all on laptops and phones, not talking, sitting together yet isolated, as though their bodies, hungry, aching for companionship, aren't there at all. The body sends its kaleidoscopic barrage of sensory information to the brain, but the way is blocked by Mark Zuckerberg and the ghost of Steve Jobs, armed to the teeth with distraction.

The thing that makes acknowledging and accepting an embodied approach to learning so difficult is that it's messy and excruciatingly difficult to verify and measure. It's Dionysian. How, exactly, can we accurately grade a student's engagement with the sense—assess their sensual experience and ability to translate that experience into well-reasoned behavior or compassion-driven action? And would we even really *want* to measure that sort of thing?

If we can accept that cognition is embodied, then education becomes an engagement with the whole spectrum of human experience—the emotional, physical, behavioral, sensual life—and we radically rewrite what it means to learn, and how to go about doing it.

This can sound alarmist; duly noted. But there seems to be a strong possibility that we are moving further away from the direct experience of the body and more toward a disconnected, disembodied culture. Our bodies are messy and imperfect and complex. The way in which we collect information about our own experience and transmute it into behavior and a sense of "I am" isn't linear or composed of a single track. It is, instead, more like the interlocking ecosystem of the forest, all tangled roots and branches and growth and decay, as much going on under the surface as above.

Not only do *we all have a body*, but *we are our body*. Our "self" is not some holographic image floating about in the ether; it is the way in which our physical forms sense, collect, and interpret experience and use that information to guide our actions. We are all muscle and tissue and bone, sure, but what a glorious reality that is.

Apollo is order, Dionysus, chaos. It's a helpful story to frame ideas about learning, but ultimately it's a simplification. Apollo may have been the god of order, but his temple at Delphi was placed over a steaming volcanic rift in the earth—order masking the chaos within. And although Dionysus may have been disorder incarnate, the god of wine and revelry names music and harmony and cultivation as attributes as well—all of which depend on some form of order.

But although the ancient Greeks may have appreciated the polyphonic complexity of the two gods—and certainly their respective cults are indicative of a robust and varied understanding of the symbolic importance of these deities—the way these two gods inform institutions today is much more binary. Apollo and the organization brigade have ruled education for a long time, and Descartes's separation of mind and body echo through to the current day. With advances in cognitive science, and the new understanding brought about by embodied cognition, it's time to revisit the wild lessons of Dionysus.

NOTES

1. Nicholas Carr, *The Glass Cage: Automation and Us* (New York: W. W. Norton, 2014).

SIXTEEN
The Jumper

There's a swimming hole near my home called Bristol Falls. It's an incredible spot, a deep gorge fed by a cold mountain river with a waterfall at one end of the long, golden-green pool. On both sides there are great ledges and blocks of rock, perfect for jumping. A slick ledge leads up to the falls. The intrepid swimmer can clamber up behind the cascade itself into a dripping open-mouthed cave hidden by the falling water and dive right through the falls. On any given day during the height of summer, dozens of people visit, sunning themselves on rocks, splashing through the shallows, piling rocks to make little pools, and leaping off the cliffs at the deep end of the pool.

In my narrowly circumscribed world, I consider it to be one of the most beautiful spots on earth. There's deep, old magic there to be sure.

I run a little outdoor adventure camp each summer; we often go to Bristol Falls. Once we get there, the kids change from squealing, rioting preteens hollering Katy Perry lyrics to something else more in tune with the world around them. They clamber over rocks, climbing up the huge slabs sticking out over the pool. They build little dams with slick, softball-sized rocks. They look for critters in the shallows. It seems they're a part of the landscape, scrabbling with both hands and feet over the tumbled rocks, hunched over and flattened out to cross foaming chutes of cold mountain stream, moving with the staccato rhythm of woodland creatures. A scurry, then a pause, more scurrying, a flipped rock. Like foraging animals.

Later, they spread their bodies out to dry on a hot rock after a cold swim. There are moments when they are so present and within themselves, toes gripping the slippery river bottom, bodies slipping through the clear water like minnows.

There's one jumping spot in particular. It juts out right next to the top of the falls. The prong of rock is like a runway; swimmers can get a head of steam and go leaping off, eighteen feet in the air, and plummet into the dark, churning waters below right at the base of the falls. Last summer, my son Finn worked up the gumption to try it. He'd jumped from boulders and rocks there and elsewhere—had already conquered the high dive at the town pool—but this is a big jump.

I stand with him, slapping mosquitoes as his face shines with excitement and he fires questions at me: "Will it hurt?" "How far out should I jump?" "Is it deep enough?" He smiles the whole time, body bobbing like a boxer's. I answer as best I can, but he isn't really listening. The questions, the joyful fear, the tension—it's all a part of the ritual.

He inhabits the moment fully, his body shining with energy.

He looks out over the edge, his face alive with the joyful anticipation of flight. He's beaming, his entire body, wet and slick as a seal's, radiating vitality. "Check for me, Dad," he says, and I do. No swimmers below. "You're good to go," I tell him. He doesn't think twice. Arms pumping, legs skipping along barely touching the ground, he dashes toward the launching spot, rockets to the end, and flings himself off. There's something admirably simian in the way he jumps, like a gibbon traveling through the canopy, his whole body part of a bend of youthful abandon, etched in the moment, legs akimbo, body thrumming like a plucked string. He's there, in that moment, in his body; he *is* his body.

It's less than an instant that he's suspended, elastically inscribing an arc as he whoops loudly. Then he's plunging into the water below.

Appendix

A Practical Guide to Embodied Education

Minimize screen time. Discontinuing the use of educational technology products and screen time in the classroom in general may disrupt the addictive feedback loop built into software platforms modeled after social media products—and make for more settled, focused students. Instead of online grading and feedback, face-to-face interactions could make a more lasting impression; positive interpersonal interactions have myriad embodied benefits.

Go outside. Outdoor time in unstructured environments (woods, fields, vacant lots, or even city parks) can possibly have a positive effect on students' global precedence, improve their memory, provide opportunities for healthy exercise, and benefit other learning experiences through bolstering embodied cognitive abilities. Plus, they'll love you for it, and a little goodwill goes a long way.

Include movement in daily lesson plans. Tai Chi, yoga, dance, acrobatics, hip hop dance, you name it: any form of embodied physical engagement will have far-reaching effects on students' learning, memory, emotions, and health. Creating lessons and schedules that include long stretches of embodied activity—particularly those that hone interoceptive, proprioceptive, and corporoceptive abilities—will improve student performance and sense of well-being. Happy and healthy students are much easier to teach than anxiety-ridden, disembodied ones.

Include mindfulness meditation exercises. Anxiety and depression—two of the leading mental health issues facing students today—can both be reduced through actively engaging in embodied practices such as mindfulness meditation. The way in which the default-mode network affects individuals with a proclivity for anxiety and depression can lead to the type of stress that exacerbates mental health issues. One possible way to ameliorate the pain of these conditions is to offer embodied practice not as a tacked-on experience in the school day, but as an integral part of learning.

Read aloud with students. Reading aloud with students is a dynamic, two-way experience. Bringing books to life through reading aloud—whether it's a group of toddlers or hungover undergrads—improves memory, increases retention, and aids in comprehension. Having students act out scenes (even in college!) and create physical representations of ideas from books increases comprehension and promotes an embodied engagement with the text.

Use actual, physical books instead of screens. Real books, rather than e-readers, reinforce the students' naturally haptic way of interacting with the world around them. Reading through a book, flipping the pages, marking progress as the wedge of pages read begins to be larger than the pages remaining—these are all physical, embodied engagements with knowledge and narrative. Reading e-readers is akin to flying over the ocean in an airplane: there's no sense of space or distance.

Have students take notes longhand with a pen or pencil. Taking notes by hand allows students to synthesize, remember, record, interpret, be concise, and analyze information—all agreed-upon goals of modern education. In addition, we know what the result will be from asking students to write by hand: solid letter-recognition skills, higher reading comprehension, improved haptic motor control. We don't yet know what the long-term effects of typing will be on embodiment, cognition, and learning.

Create some meditation exercises to increase critical distance, helping students and teachers be less likely to get stressed by critiques. Students who are stressed by feedback would benefit from gaining some distance from and perspective on their studies. The best way to do this is to work through mindfulness exercises, whether it's meditation or aikido, to create an embodied sense of "centeredness" from which to absorb, process, and respond to critical feedback. It's not a bad idea for teachers who suffer from faculty meeting burnout, either.

Include mindfulness exercises to encourage positive student behavior. Emotional regulation is tied directly to interoceptive awareness. Students who fly off the handle, respond inappropriately, or have trouble reading social cues appropriately (not taking into account neurodiversity, a subject outside the purview of this book) would benefit from mindfulness practice. Mindfulness meditation and similar exercises help kids "tune in" to the frequencies of their bodies, thus developing an awareness of how they feel. And how they feel is tied to how they act.

Educate, promote, and model good eating habits (wait until you get home to snarf that whole box of Fruit Loops). What students eat can have far-reaching effects on not just fitness but mental-health issues such as anxiety and depression, as well as have ramifications for cognitive abilities. In states across the nation, low-income students qualify for free or reduced-cost meals; this is one of the greatest reforms public education has ever achieved. For educators, keeping in mind the manner in which diet can alter student performance and ability is vital. We all love our Ben and Jerry's and Slurpees, but being intentional about what we put in our bodies and how we educate our students around diet is directly tied to how we all feel, act, and learn.

Find a way to have white people embody the experience of minorities. Diversity work needs to be carried out in an embodied fashion for both students and educators. This enterprise is fraught with risk, obviously. It has to be taken seriously and done intentionally. But for any real progress to be made in diversity, equity, and inclusion issues, white teachers and students have to engage in exercises in which they can embody some of the discrimination felt by students and teachers of color. The same goes for work on gender and LGBQT issues. Only through active embodiment exercises can the effect of bias be felt, internalized, and embodied by white individuals.

Don't assume that everyone interprets text the same, and remember that this has nothing to do with academic ability. As far as biological sex goes, men and women have different levels of reading comprehension depending on the text. In addition, they experience body language differently. Thus, grading or assessing or viewing the playing field as flat—equating the experience of the two sexes—is a faulty approach. Context matters and so does biological individuality when it comes to learning and memory.

Integrate movement into any and all learning experiences possible. Creating "lifelong learners" needs to incorporate embodiment work to truly benefit students. Generally, humans become less embodied, less attuned to their sensorimotor perceptions, as they grow older. This can have a negative effect on learning. Any educational institution that promises the development of "lifelong learners" or offers adult education programs would do well to consider creating programs that include direct engagement with embodiment principles: movement, mindfulness, and exercise.

Get in touch with your own sense of embodiment. Students follow the examples of their peers and, to a lesser extent, their teachers. How can educators model embodied awareness principles? How can a teacher demonstrate that being in one's body is a necessary component—central, I'd say—to the project of personal development and learning? Not everyone is going to be a yogi or an ultra-runner, but modeling awareness and engagement with the body sends the message to students that, yes, your body and how it feels matters. And that's something we have a least a bit of control over.

Bibliography

Abram, David. *The Spell of the Sensuous: Perception and Language in a More-than-Human World.* New York: Vintage, 1996.

Adam, Hajo, and Adam D. Galinsky. "Enclothed Cognition." *Journal of Experimental Social Psychology* 48, no. 4 (2012): 918–25, doi:10.1016/j.jesp.2012.02.008.

"Adopted Romanian Orphans 'Still Suffering in Adulthood.'" *BBC News*, BBC, February 23, 2017. www.bbc.com/news/health-39055704.

Agusti, Ana, et al. "Interplay Between the Gut-Brain Axis, Obesity and Cognitive Function." *Frontiers*, Feb. 26, 2018. https://www.frontiersin.org/articles/10.3389/fnins.2018.00155/full.

Anderson, Steven W., et al. "The Earliest Behavioral Expression of Focal Damage to Human Prefrontal Cortex." *Egyptian Journal of Medical Human Genetics* March 6, 2008. www.sciencedirect.com/science/article/pii/S0010945208705082.

Avenanti, Alessio, et al. "Racial Bias Reduces Empathic Sensorimotor Resonance with Other-Race Pain." *Current Biology.* 20, no. 11 (2010): 1018–22. doi:10.1016/j.cub.2010.03.071.

Banakou, Domna, et al. "Virtual Embodiment of White People in a Black Virtual Body Leads to a Sustained Reduction in Their Implicit Racial Bias." *Frontiers in Human Neuroscience*, Nov. 29, 2016.

Bechara, Antoine, et al. "Deciding Advantageously before Knowing the Advantageous Strategy." *Science*, Feb. 28, 1997. science.sciencemag.org/content/275/5304/1293.full.

Benjamin, R. and Schwanenflugel, P. 2010. Text Complexity and Oral Reading Prosody in Young Readers. *Reading Research Quarterly*, 45(4), 388–404.

Bergen, Benjamin K. *Louder than Words: the New Science of How the Mind Makes Meaning.* Basic Books, 2012.

Berwid, Olga G., and Jeffrey M. Halperin. "Emerging Support for a Role of Exercise in Attention-Deficit/Hyperactivity Disorder Intervention Planning." *Current Psychiatry Reports*, Oct. 2012. www.ncbi.nlm.nih.gov/pmc/articles/PMC3724411/.

Bloom, Stephen G. "Lesson of a Lifetime." *Smithsonian.com*, Sep. 1, 2005. www.smithsonianmag.com/science-nature/lesson-of-a-lifetime-72754306/.

Bregman, Peter. "Diversity Training Doesn't Work." *Psychology Today*, March 12, 2012. www.psychologytoday.com/us/blog/how-we-work/201203/diversity-training-doesnt-work.

Brown, Richard E. "Hebb and Cattell: The Genesis of the Theory of Fluid and Crystallized Intelligence." *Frontiers in Human Neuroscience*, vol. 10 (2016), doi:10.3389/fnhum.2016.00606.

Buckner, Randy L., et al. "Opportunities and Limitations of Intrinsic Functional Connectivity MRI." *Nature News*, June 25, 2013. www.nature.com/articles/nn.3423.

Cahill, Larry. "Why Sex Matters for Neuroscience." *Nature News*, May 10, 2006. www.nature.com/articles/nrn1909.

Calvino, Italo. *If on a Winter's Night a Traveler*. Translated by William Weaver. New York: Harcourt Brace & Company, 1999.
ClassicSoulRadio. YouTube, Feb. 1, 2012. https://www.youtube.com/watch?v=lODBVM802H8.
Claxton, Guy. *Intelligence in the Flesh*. New Haven, CT: Yale University Press, 2016.
Cleaver, Samantha, and Munro Richardson. *Read with Me: Engaging Your Young Child in Active Reading*. Rowman & Littlefield, 2019.
"College Students' Mental Health Is a Growing Concern, Study Finds." *Monitor on Psychology*, June 2013. www.apa.org/monitor/2013/06/college-students.aspx.
Coryell, William, et al. "Subgenual Prefrontal Cortex Volumes in Major Depressive Disorder and Schizophrenia: Diagnostic Specificity and Prognostic Implications." *American Journal of Psychiatry* (September 2005). ajp.psychiatryonline.org/doi/pdf/10.1176/appi.ajp.162.9.1706.
Cunningham, Anne E., and Keith E. Stanovich. "Early Spelling Acquisition: Writing Beats the Computer." *Journal of Educational Psychology* 82, no. 1 (1990): 159–62. doi:10.1037//0022-0663.82.1.159.
Damasio, Antonio R. *Descartes' Error: Emotion, Reason and the Human Brain*. New York: Vintage, 2006.
Damasio, Antonio R. *Looking for Spinoza: Joy, Sorrow, and the Feeling Brain*. New York: Vintage, 2004.
Damasio, Antonio R. "The Somatic Marker Hypothesis and the Possible Functions of the Prefrontal Cortex." *The Prefrontal CortexExecutive and Cognitive Functions*, Apr. 1998. doi:10.1093/acprof:oso/9780198524410.003.0004, pp. 36–50.
Derntl, B., et al. "General and Specific Responsiveness of the Amygdala during Explicit Emotion Recognition in Females and Males." *BMC Neuroscience* (Aug. 4, 2009).
Descartes, Rene. *A Discourse on Method; Meditations on the First Philosophy; Principles of Philosophy*. London, UK: Forgotten Books, 2015.
Dewey, John. *Democracy and Education*. SMK Books, 2018.
Dunn, B. D., et al. "Listening to Your Heart. How Interoception Shapes Emotion Experience and Intuitive Decision Making." *Current Neurology and Neuroscience Reports* (Dec. 2010). www.ncbi.nlm.nih.gov/pubmed/21106893.
Dunn, Barnaby D., et al. "Gut Feelings and the Reaction to Perceived Inequity: The Interplay between Bodily Responses, Regulation, and Perception Shapes the Rejection of Unfair Offers on the Ultimatum Game." *Cognitive, Affective, and Behavioral Neuroscience*, (May 23, 2012). link.springer.com/article/10.3758/s13415-012-0092-z.
Farb, Norman, et al. "Interoception, Contemplative Practice, and Health." *Frontiers* (May 22, 2015). www.frontiersin.org/articles/10.3389/fpsyg.2015.00763/full.
Forgiarini, Matteo, et al. "Racism and the Empathy for Pain on Our Skin." *Frontiers in Psychology*, May 23, 2011, https://www.ncbi.nlm.nih.gov/pmc/articles/PMC3108582.
Fung, Thomas C., et al. "Intestinal Serotonin and Fluoxetine Exposure Modulate Bacterial Colonization in the Gut." *Nature Microbiology*, Feb. 2019, doi:10.1038/s41564-019-0540-4.
Gapenne, Olivier. "The Co-Constitution of the Self and the World: Action and Proprioceptive Coupling." *Frontiers in Psychology*, June 12, 2014, https://www.ncbi.nlm.nih.gov/pmc/articles/PMC4054590/.
Genschow, Oliver, et al. "Mimicking and Anticipating Others' Actions Is Linked to Social Information Processing." *Plos One*, vol. 13, no. 3 (2018). doi:10.1371/journal.pone.0193743.
Gilbert, Elizabeth. *Big Magic: Creative Living beyond Fear*. New York: Riverhead, 2015.

Glenberg, Arthur M., et al. "Gender, Emotion, and the Embodiment of Language Comprehension," *Emotion Review*. 1, no. 2 (2009): 151–61. doi:10.1177/1754073908100440.

Glenberg, Arthur M., et al. "Use-Induced Motor Plasticity Affects the Processing of Abstract and Concrete Language." *Current Biology*, vol. 18, no. 7 (2008). doi:10.1016/j.cub.2008.02.036.

Godden, D. R., and A. D. Baddeley. "Context-Dependent Memory in Two Natural Environments: On Land and Underwater." *British Journal of Psychology*, vol. 66, no. 3 (1975), pp. 325–331, doi:10.1111/j.2044-8295.1975.tb01468.x.

Gogolla, Nadine. "The Insular Cortex." *Current Neurology and Neuroscience Reports* (June 19, 2017). www.ncbi.nlm.nih.gov/pubmed/28633023.

Hamilton, J. Paul, et al. "Depressive Rumination, the Default-Mode Network, and the Dark Matter of Clinical Neuroscience." *Biological Psychiatry*, Aug. 15, 2015. https://www.ncbi.nlm.nih.gov/pubmed/25861700.

Hanh, Thich Nhat. *You Are Here*. Boulder, CO: Shambhala Publications, 2009.

Husserl, Edmund, and William P. Alston. *The Idea of Phenomenology*. Dordrecht, Netherlands: Kluwer Academic, 2010.

Husserl, Edmund. *Ideas: General Introduction to Pure Phenomenology*. New York: Routledge, 2015.

Imlach, Abbie-Rose, et al. "Age Is No Barrier: Predictors of Academic Success in Older Learners." *Nature News* (Nov. 15, 2017). www.nature.com/articles/s41539-017-0014-5.

Inuggi, Alberto, et al. "Brain Functional Connectivity Changes in Children That Differ in Impulsivity Temperamental Trait." *Advances in Pediatrics* (2014). www.ncbi.nlm.nih.gov/pmc/articles/PMC4018550/.

Isaacson, Walter. *Einstein: His Life and Universe*. New York: Simon & Schuster, 2008.

James, Karin H., and Isabel Gauthier. "Letter Processing Automatically Recruits a Sensory–Motor Brain Network." *Neuropsychologia* 44, no. 14 (2006): 2937–49. doi:10.1016/j.neuropsychologia.2006.06.026.

Jenkins, Tiara, and Jessica Yarmosky. "Dr. Seuss Books Can Be Racist, But Students Keep Reading Them." *NPR*, Feb. 26, 2019, https://www.npr.org/sections/codeswitch/2019/02/26/695966537/classic-books-are-full-of-problems-why-cant-we-put-them-down.

Johnson, Brad. *Learning on Your Feet: Incorporating Physical Activity into the K–8 Classroom*. New York: Routledge, Taylor and Francis, 2016.

Johnson, Katerina V. A., and Kevin R. Foster. "Why Does the Microbiome Affect Behaviour?" *Nature Reviews Microbiology* 16, no. 10 (2018): 647–55. doi:10.1038/s41579-018-0014-3.

Johnson, Mark. *The Meaning of the Body: Aesthetics of Human Understanding*. Chicago: University of Chicago Press, 2012.

Johnston, Bret Anthony. "Don't Write What You Know." *Atlantic*, Feb. 19, 2014, https://www.theatlantic.com/magazine/archive/2011/08/dont-write-what-you-know/308576/.

Kabat-Zinn, Jon. *Coming to Our Senses: Healing Ourselves and the World through Mindfulness*. New York: Hyperion, 2005.

Kabat-Zinn, Jon. *Full Catastrophe Living: Using the Wisdom of Your Body and Mind to Face Stress, Pain, and Illness*. New York: Bantam. 2013.

Kabat-Zinn, Jon. *Wherever You Go, There You Are: Mindfulness Meditation in Everyday Life*. New York: Hyperion, 2005.

Kalanthroff, E., et al. "Spatial Processing in Adults with Attention Deficit Hyperactivity Disorder." *Neuropsychology* (September 2013). www.ncbi.nlm.nih.gov/pubmed/23937479.

Kalev, A., Kelley, E., and F. Dubbin. "Diversity Management in Corporate America" *Contexts*, vol. 6, no. 4, 21–27.

Kant, Immanuel, and Allen W. Wood. *Basic Writings of Kant*. New York: Modern Library Classics, 2001.

King, Stephen. *On Writing: A Memoir of the Craft*. New York: Scribner, 2000.

Kress, Gunther R. *Literacy in the New Media Age*. New York: Routledge, 2010.

Lakoff, George, and Mark Johnson. *Philosophy in the Flesh: The Embodied Mind and Its Challenge to Western Thought*. New York: Basic, 1999.

Lengel, Traci, and Mike Kuczala. *The Kinesthetic Classroom: Teaching and Learning through Movement*. Thousand Oaks, CA: Corwin, 2010.

Lepecki André. *Of the Presence of the Body: Essays on Dance and Performance Theory*. Middletown, CT: Wesleyan University Press, 2006.

Li, Hongjie. "Diet, Gut Microbiota and Obesity." *Journal of Nutritional Health & Food Science* 3, no. 4 (2015): 1–6. doi:10.15226/jnhfs.2015.00150.

Longcamp, Marieke, et al. "Remembering the Orientation of Newly Learned Characters Depends on the Associated Writing Knowledge: A Comparison between Handwriting and Typing." *Human Movement Science* 25, no. 4–5 (2006): 646–56. doi:10.1016/j.humov.2006.07.007.

Longcamp, Marieke, et al. "Visual Presentation of Single Letters Activates a Premotor Area Involved in Writing." *NeuroImage* 19, no. 4 (2003): 1492–500. doi:10.1016/s1053-8119(03)00088-0.

MacLeod, C. M. "The production effect: delineation of a phenomenon." National Center for Biotechnology Information (2019). www.ncbi.nlm.nih.gov/pubmed/20438265.

Mangen, Anne, Jean-Luc Velay. "Digitizing Literacy: Reflections on the Haptics of Writing." *IntechOpen* (Apr. 1, 2010), https://www.intechopen.com/books/advances-in-haptics/digitizing-literacy-reflections-on-the-haptics-of-writing.

Mantis, N. J., et al. "Secretory IgA's Complex Roles in Immunity and Mucosal Homeostasis in the Gut." *Nature News* (Oct. 5, 2011). www.nature.com/articles/mi201141.

Marasco, Paul D. "Using Proprioception to Get a Better Grasp on Embodiment." *Journal of Physiology* (Jan. 15, 2018). www.ncbi.nlm.nih.gov/pubmed/29194626.

Marjan, Laal, and Peyman Salamati. "Lifelong Learning: Why Do We Need It?" *NeuroImage* (Jan. 13, 2012). www.sciencedirect.com/science/article/pii/S1877042811030023.

Marsh, Sarah. "Xanax Misuse: Doctors Warn of 'Emerging Crisis' as UK Sales Rise." *Guardian* (Feb. 5, 2018), https://www.theguardian.com/society/2018/feb/05/xanax-misuse-uk-dark-web-sales-health.

Meyer, Robinson. "Were Our Ancestors Sleeping in Trees 3 Million Years Ago?" *Atlantic* (July 6, 2018). www.theatlantic.com/science/archive/2018/07/australopithecus-foot/564422/.

Mueller, Pam M., and Daniel M. Oppenheimer. "The Pen Is Mightier Than the Keyboard: Advantages of Longhand Over Laptop Note Taking." *Psychological Research* (April 23, 2014). journals.sagepub.com/doi/abs/10.1177/0956797614524581?journalCode=pssa.

Newen, Albert. "The Embodied Self, the Pattern Theory of Self, and the Predictive Mind." *Frontiers in Psychology* (Nov. 23, 2018), www.ncbi.nlm.nih.gov/pmc/articles/PMC6265368/.

Okumura, Ryu, and Kiyoshi Takeda. "Roles of Intestinal Epithelial Cells in the Maintenance of Gut Homeostasis." *Experimental & Molecular Medicine* (May 26, 2017). www.ncbi.nlm.nih.gov/pmc/articles/PMC5454438/.

Overton, Willis F., et al. "Developmental Perspectives on Embodiment and Consciousness." *Psychology*, 2013.

Pagis, Michael. "Embodied Self-Reflexivity." *Social Psychology Quarterly* (September 2009).

Paulus, Martin P. "Decision-Making Dysfunctions in Psychiatry-Altered Homeostatic Processing?" *Science* (Oct. 26, 2007). science.sciencemag.org/content/318/5850/602.

Peck, Tabitha C., et al. "Putting Yourself in the Skin of a Black Avatar Reduces Implicit Racial Bias." *Consciousness and Cognition* 22, no. 3 (2013): 779–87. doi:10.1016/j.concog.2013.04.016.

Perlez, Jane. "Romanian 'Orphans': Prisoners of Their Cribs." *New York Times* (Mar. 25, 1996). https://www.nytimes.com/1996/03/25/world/romanian-orphans-prisoners-of-their-cribs.html.

Pomerantz, J. R. "Global and Local Precedence: Selective Attention in Form and Motion Perception." *Journal of Experimental Psychology* (Dec. 1983). www.ncbi.nlm.nih.gov/pubmed/6229597.

Pontzer, Herman. "Overview of Hominin Evolution." *Nature News* (2012). www.nature.com/scitable/knowledge/library/overview-of-hominin-evolution-89010983.

Praechter, Carrie. "Power, Knowledge and Embodiment in Communities of Sex/Gender Practice." *NeuroImage*(Nov. 21, 2005). www.sciencedirect.com/science/article/abs/pii/S0277539505000890.

Price, Cynthia J., and Carole Hooven. "Interoceptive Awareness Skills for Emotion Regulation: Theory and Approach of Mindful Awareness in Body-Oriented Therapy (MABT)." *Frontiers in Psychology*, vol. 9 (2018). doi:10.3389/fpsyg.2018.00798.

Proctor, Cicely, et al. "Diet, Gut Microbiota and Cognition." *Metabolic Brain Disease* (February 2017). www.ncbi.nlm.nih.gov/pubmed/27709426.

Prum, Richard O. *The Evolution of Beauty: How Darwin's Forgotten Theory of Mate Choice Shapes the Animal World—and Us.* New York: Anchor Books, 2018.

"Race and Empathy Matter on Neural Level." *ScienceDaily*, (Apr. 27, 2010). www.sciencedaily.com/releases/2010/04/100426182002.htm.

Romero, and Mary. "Reflections on 'The Department Is Very Male, Very White, Very Old, and Very Conservative': The Functioning of the Hidden Curriculum in Graduate Sociology Departments." *OUP Academic* (Apr. 8, 2017). https://academic.oup.com/socpro/article/64/2/212/3231922.

Salzberg, Sharon. *Real Happiness: Learn the Power of Meditation: A 28-Day Program.* New York: Workman, 2010.

Salzberg, Sharon. *Real Happiness at Work: Meditations for Accomplishment, Achievement, and Peace.* New York: Workman, 2015.

Salzberg, Sharon. *Real Love: The Art of Mindful Connection.* New York: Flatiron Books, 2017.

Sapolsky, Robert. *Behave: The Biology of Humans at Our Best and Worst.* New York: Vintage, 2018.

Schmalz, Julia. "Facing Anxiety: Students share how they cope and how campuses can help," *Chronicle of Higher Education* (Dec. 11, 2017). www.chronicle.com/article/Facing-Anxiety/241968.

Seidler, Guenter H., and Frank E. Wagner. "Comparing the Efficacy of EMDR and Trauma-Focused Cognitive-Behavioral Therapy in the Treatment of PTSD: A Meta-Analytic Study." *Psychological Medicine* 36, no. 11 (Nov. 2006): 1515–522. www.cambridge.org/core/journals/psychological-medicine/article/comparing-the-efficacy-of-emdr-and-traumafocused-cognitivebehavioral-therapy-in-the-treatment-of-ptsd-a-metaanalytic-study/F4CD874AE857F3D08AA2EF1181F0098B.

Servaas, Michelle Nadine, et al. "The Effect of Criticism on Functional Brain Connectivity and Associations with Neuroticism." *British Journal of Educational Psychology* (Oct. 12, 2017). www.rug.nl/research/portal/en/publications/the-effect-of-criticism-on-functional-brain-connectivity-and-associations-with-neuroticism(62013cbb-6a05-416c-839c-8640f052b7b9).html.

Shapiro, Francine. "The Role of Eye Movement Desensitization and Reprocessing (EMDR) Therapy in Medicine: Addressing the Psychological and Physical Symptoms Stemming from Adverse Life Experiences." *Permanente Journal* (2014). https://www.ncbi.nlm.nih.gov/pmc/articles/PMC3951033/.

Shapiro, Lawrence A. *Embodied Cognition*. New York: Routledge, 2011.

Sheets-Johnstone, Maxine. *The Primacy of Movement*. Amsterdam: John Benjamins, 1999.

Shenk, Joshua Wolf. *Powers of Two: Finding the Essence of Innovation in Creative Pairs*. London: John Murray, 2015.

Shoecraft, Stacey. *Teaching through Movement: Setting Up Your Kinesthetic Classroom*. CreateSpace Independent Publishing Platform, 2015.

Silver, Steven M., et al. "EMDR Therapy Following the 9/11 Terrorist Attacks: A Community-Based Intervention Project in New York City." *International Journal of Stress Management* (2005).

Singer, Tania, et al. "A Common Role of Insula in Feelings, Empathy and Uncertainty." *NeuroImage* (July 28, 2009). www.sciencedirect.com/science/article/pii/S1364661309001351.

Slepian, Michael L., et al. "Tough and Tender." *Psychological Science* 22, no. 1 (2010): 26–28. doi:10.1177/0956797610390388.

Smith, Peter Andrey. "The Tantalizing Links between Gut Microbes and the Brain." *Nature News* (Oct. 14, 2015). www.nature.com/news/the-tantalizing-links-between-gut-microbes-and-the-brain-1.18557.

Smith, Zadie. "Dance Lessons for Writers." *Guardian*, Oct. 29, 2016. Retrieved from https://www.theguardian.com/books/2016/oct/29/zadie-smith-what-beyonce-taught-me.

Sokolov, Arseny A., et al. "Gender Affects Body Language Reading." *Frontiers in Psychology*, vol. 2 (Feb. 2, 2011). doi:10.3389/fpsyg.2011.00016.

Spielberg, Jeffrey M., et al. "Prefrontal Cortex, Emotion, and Approach/Withdrawal Motivation." *Advances in Pediatrics* (Jan. 1, 2008). www.ncbi.nlm.nih.gov/pmc/articles/PMC2889703/.

Spunt, Robert P, and Matthew D Lieberman. "An Integrative Model of the Neural Systems Supporting the Comprehension of Observed Emotional Behavior." *NeuroImage* (Oct. 13, 2011). www.sciencedirect.com/science/article/pii/S1053811911011669.

Stutz, Aaron J. "Embodied Niche Construction in the Hominin Lineage: Semiotic Structure and Sustained Attention in Human Embodied Cognition." *Frontiers in Psychology* (Aug. 1, 2014). https://www.ncbi.nlm.nih.gov/pmc/articles/PMC4117988/

Svoboda, Elizabeth. "Can Microbes Encourage Altruism?" *Quanta Magazine* (June 29, 2017). www.quantamagazine.org/can-microbes-encourage-altruism-20170629/.

Tomizawa, Kazuhito, et al. "Oxytocin Improves Long-Lasting Spatial Memory during Motherhood through MAP Kinase Cascade." *Nature Neuroscience*, vol. 6, no. 4 (2003), pp. 384–90. doi:10.1038/nn1023.

"Touch Can Produce Detailed, Lasting Memories." *Neuroscience News* (Nov. 27, 2018), https://neurosciencenews.com/touch-memory-10262/.

Tussyadiah, Iis P., et al. "Embodiment of Wearable Augmented Reality Technology in Tourism Experiences." *Journal of Travel Research* (2017). journals.sagepub.com/doi/abs/10.1177/0047287517709090.

van Elk, Michiel, et al. "Embodied Language Comprehension Requires an Enactivist Paradigm of Cognition." *Frontiers in Psychology*, Dec. 27, 2010. https://www.ncbi.nlm.nih.gov/pmc/articles/PMC3153838/.

Waal, Frans de. *Mama's Last Hug: Animal Emotions and What They Tell Us about Ourselves*. New York: W. W. Norton, 2019.

"Watching Others Makes People Overconfident in Their Own Abilities." *Association for Psychological Science (APS)*. https://www.psychologicalscience.org/news/releases/watching-others-makes-people-overconfident-in-their-own-abilities.html.

Vaughn, Sharon, et al. "Early Spelling Acquisition: Does Writing Really Beat the Computer?" *Learning Disability Quarterly* 15, no. 3 (1992): 223–28. doi:10.2307/1510245.

Wehrenberg, Margaret. "Rumination: A Problem in Anxiety and Depression." *Psychology Today* (April 20, 2016). www.psychologytoday.com/us/blog/depression-management-techniques/201604/rumination-problem-in-anxiety-and-depression.

Weil, Elizabeth. "Why He Kayaked across the Atlantic at 70 (for the Third Time)." *New York Times*, March 22, 2018. www.nytimes.com/interactive/2018/03/22/magazine/voyages-kayaking-across-ocean-at-70.html.

Weiss, Bari. "Try to Keep Up with Australia's Fastest 92-Year-Old Woman." *New York Times*, Jan. 4, 2019. www.nytimes.com/2019/01/04/opinion/australia-racewalking-aging-healthy.html.

Wilson, Andrew D., and Sabrina Golonka. "Embodied Cognition Is Not What You Think It Is." *Frontiers in Psychology*, vol. 4, Feb. 12, 2013. doi:10.3389/fpsyg.2013.00058.

Wilson, Frank. *The Hand: How Its Use Shapes the Brain, Language, and Human Culture*. New York: Pantheon, 1998.

Wilson, Sandra A., et al. "Stress Management with Law Enforcement Personnel: A Controlled Outcome Study of EMDR Versus a Traditional Stress Management Program." New York: Kluwer Academic-Plenum Publishers, July 2001. https://link.springer.com/article/10.1023/A:1011366408693.

Zahavi, Dan. *Self and Other: Exploring Subjectivity, Empathy, and Shame*. New York: Oxford University Press, 2016.

About the Author

Erik Shonstrom is the author of *Wild Curiosity* and *The Indoor Epidemic*, and has written for *The Chronicle of Higher Education*, *The Los Angeles Review of Books*, and *Education Week*. Erik received his master of fine arts in writing from Bennington College. He teaches at Champlain College in Vermont, where he lives with his family.

www.ingramcontent.com/pod-product-compliance
Lightning Source LLC
Chambersburg PA
CBHW030138240426
43672CB00005B/179